Mind and Life

Columbia Series in Science and Religion

The Columbia Series in Science and Religion

The Columbia Series in Science and Religion is sponsored by the Center for the Study of Science and Religion (CSSR) at Columbia University. It is a forum for the examination of issues that lie at the boundary of these two complementary ways of comprehending the world and our place in it. By examining the intersections between one or more of the sciences and one or more religions, the CSSR hopes to stimulate dialogue and encourage understanding.

Robert Pollack, *The Faith of Biology and the Biology of Faith*

B. Alan Wallace, ed., *Buddhism and Science: Breaking New Ground*

Lisa Sideris, *Environmental Ethics, Ecological Theory, and Natural Selection: Suffering and Responsibility*

Wayne Proudfoot, ed., *William James and a Science of Religions: Reexperiencing the Varieties of Religious Experience*

Mortimer Ostow, *Spirit, Mind, and Brain: A Psychoanalytic Examination of Spirituality and Religion*

B. Alan Wallace, *Contemplative Science: Where Buddhism and Neuroscience Converge*

Philip Clayton and Jim Schaal, editors, *Practicing Science, Living Faith: Interviews with Twelve Scientists*

B. Alan Wallace, *Hidden Dimensions: The Unification of Physics and Consciousness*

Mind and Life

DISCUSSIONS WITH THE **DALAI LAMA**
ON THE **NATURE OF REALITY**

PIER LUIGI LUISI

WITH THE ASSISTANCE OF ZARA HOUSHMAND

Columbia University Press   New York

Columbia University Press

Publishers Since 1893

New York   Chichester, West Sussex

Copyright © 2009 Mind and Life Institute

All rights reserved

Library of Congress Cataloging-in-Publication Data

Luisi, P. L.

Mind and life : discussions with the Dalai Lama on the nature of reality / Pier Luigi Luisi
   with the assistance of Zara Houshmand.

   p.   cm. — (The Columbia series in science and religion)

Includes bibliographical references and index.

ISBN 978-0-231-14550-3 (cloth : alk. paper)

1. Matter—Religious aspects—Buddhism. 2. Life—Religious aspects—Buddhism. 3.
   Biology—Religious aspects—Buddhism. 4. Philosophy, Buddhist. I. Houshmand, Zara.
   II. Title. III. Series.

BQ4570.M37L85   2009

294.3'3657—dc22

   2008006773

Columbia University Press books are printed on permanent and durable acid-free paper.
This book was printed on paper with recycled content.

Printed in the United States of America

c 10 9 8 7 6 5 4 3 2 1

References to Internet Web sites (URLs) were accurate at the time of writing. Neither the
author nor Columbia University Press is responsible for URLs that may have expired or
changed since the manuscript was prepared.

Dedicated to the memory of Francisco Varela

From a previous Mind and Life meeting: His Holiness the Dalai Lama greeting the late Francisco Varela, to whom this book is dedicated.

Contents

Acknowledgments

I WISH TO EXPRESS my gratitude to Dr. Egbert Asshauer, Michel Bitbol, and Alan Wallace for fruitful discussion and criticism. The photos in the book come with permission from the personal archives of Carey Linde, Matthieu Ricard, Pier Luigi Luisi, and the Mind and Life Institute. Thanks are also due to Angelo Merante at the University of Rome 3 for the photo editing. Scientific illustrations are taken either from the Internet, with modifications; or from the book by P. L. Luisi, *The Emergence of Life, from Chemical Origins to Synthetic Biology* (New York: Cambridge University Press, 2006), with modifications.

Mind and Life

Introduction

THE DALAI LAMA ENTERED the room, whispering greetings one by one to familiar faces among the guests who were standing to meet him. He stopped for a special greeting for the only child present, Francisco Varela's son Gabo, and tousled his hair. Then he settled into his seat. Our host in his own home, he seemed completely at ease.

A view from McLeod Ganj, the village near Dharamsala where the Mind and Life conferences are held.

Thupten Jinpa (left) and Alan Wallace.

With him in an informal circle around the low coffee table were other representatives of two great knowledge traditions, Western science and Buddhism. Biologists Ursula Goodenough and Eric Lander would have their turns in the "hot seat," as the presenter's chair had come to be known by alumni of previous Mind and Life meetings. Later I would take that seat to unfold the molecular complexities of the origins of life, and physicist Steven Chu would have his turn to explore even more fundamental underpinnings of the scientific story. Our scientific coordinator, physicist Arthur Zajonc, brought a graceful and articulate vigilance to the role of moderator. He was our navigator, tracking progress from moment to moment on the ambitious route we had mapped in advance. As the terrain shifted unexpectedly, more intense planning sessions would continue daily over meals at the modest guesthouse that was our headquarters. The philosopher Michel Bitbol also had a special responsibility for keeping his eye on the big picture. It was he who was charged with examining the foundational assumptions on which our modern theories of matter and life depend—ideas so deeply ingrained in our thinking that we are often blind to their influence.

In the Buddhist camp, the two interpreters sitting close beside the Dalai Lama had perhaps the most demanding task of all. The Dalai Lama's command of English is substantial after many years of dealing with the West, but we would be grappling with concepts from both science and Buddhism that are highly technical and embedded in systems of thought that do not

His Holiness the Karmapa (left) and Matthieu Ricard.

map neatly onto alien worlds. Alan Wallace and Thupten Jinpa would serve not merely as translators but as cultural liaisons, ready to unpack a single term with a minilecture in English or Tibetan at a moment's notice, ever watchful for hidden rocks on our path.

The Venerable Matthieu Ricard, who often serves as the Dalai Lama's French interpreter, was another vital player. Before taking vows as a monk, Matthieu had been a brilliant young scientist specializing in cellular genetics, and his ability to straddle these two worlds had led him to a pivotal role in the Mind and Life Institute's research projects as well as its conferences.

One unusual guest was also with us at the table—His Holiness the Karmapa. The Dalai Lama had made it clear that he wanted the eighteen-year-old monk who had recently fled from China close beside him for this meeting.

Adam Engle, chairman and cofounder (with Francisco Varela) of the Mind and Life Institute, completed the group, looking very pleased at the stellar array of intellectual power he had helped to assemble. Many months of work had gone into the logistical preparation for this meeting, and now he could relax and enjoy the fruits of his labor.

Surrounding the inner circle were two more rows, a total of about sixty people. The room was more crowded than in previous years largely because of the presence of more than twenty Tibetan monks. The Dalai Lama was eager that this dialogue between Buddhism and science, which had held his own attention for well over a decade now, should begin to ripple more widely through the monastic community. Huddled immediately be-

The Venerable Amchok Rinpoche, the former head of the Library of Tibetan Works and Archives.

hind the Dalai Lama, a handful of the most senior monks sat on the edge of their seats, constantly leaning toward him. He would turn to them often, in particular to the Venerable Amchok Rinpoche, then the director of the Library of Tibetan Works and Archives, to discuss points that came up in the dialogue.

These same scholars had attended previous Mind and Life conferences, but for the ranks of younger monks the break from monastic routine was new—both a field trip into Western science and an awe-inspiring week in the Dalai Lama's presence. Additional translators planted among them were whispering softly in Tibetan, and a special session had been arranged for the young monks to question the scientists directly. Now their tightly ranked rows of shiny shaved heads seemed fiercely intent, trying not to lose a word that passed; but several times a day, this intense seriousness was shaken by a burst of laughter that passed through the human cluster like a beautiful, friendly wind.

We were all waiting for the Dalai Lama's opening words. When he began, the transparent play of expressions on his face was constantly changing—eyebrows now raised in attentive curiosity, now lowered in a brief frown of intellectual perturbation—but always it resolved to the embracing good humor of a warm smile.

"I would like to extend my greetings to all of you and also welcome you here. Many of you are already familiar with this place and our conferences. This time we are welcoming more people, and especially more monks. In the last two or three years, we have started science lessons for selected monks in South India, and already there are signs of growing interest among monk scholars. Of course, sometimes they are a little orthodox. But fortunately it seems they are beginning to realize the importance of the knowledge of modern science." He gestured to the young Karmapa. "Our guest has already started English lessons, so we can examine him to see how his English is progressing. Of course his major task is the study of Buddha Dharma. So anyway, everybody is most welcome."

A shadow passed over the Dalai Lama's face as he turned to the subject of September 11th, and he spoke very slowly. "Since the tragedy more than one year ago in America, I think here in this place we enjoy a more or less safe atmosphere. There's not much change here. So I'm very happy to have this gathering."

He then asked an attendant to hand him a framed photograph of Francisco Varela, one of the founders of the Mind and Life Institute, who had recently passed away. He placed the photograph on the table in the center of the group. "At the beginning I would like to remember our dear friend, the late Francisco Varela, who was instrumental in initiating this series of dialogues right from the beginning. Now he is no longer with us, not only here, but no longer on this planet. Whether he is reincarnated somewhere, no one knows. But here in this physical world, he is no longer with us. It's sad. He was not only a great scientist but also a very nice person. So certainly we should remember him.

"Also, another scientist, Dr. Robert Livingston, is no longer on this planet. Right from the beginning a very close feeling developed between us, and I consider him as a teacher in certain fields of science. The late David Bohm is also no longer with us. He was a great scientist, and again, such a nice person and very humble. I think of another great scientist, the physicist Professor von Weizsäcker, who still is on this planet, though he is ninety years old now. Next week when I will be in Munich I am going to meet him. I just wanted to express the memory of our old friends.

"Of course many of you already know the purpose of our conference. But some may not know, especially some of the Tibetans and newcomers here, so I would like to say a few words about our goals and the reasons why we hold this meeting. When I met with Adam Engle two days ago, he expressed some concern as to how the meetings could continue now that Francisco Varela is no longer with us. It was mainly because of him that our meetings were possible.

"I told Adam that the reason we have had this dialogue between Buddhists and scientists for quite some time is not just because a few friends

Adam Engle, the present President of the Mind and Life Institute, and a view of the conference hall.

do this for our own interest. These discussions have developed because we have certain clear objectives. Therefore we have to carry on whether a particular individual is there or not. Moreover, at the beginning there were very few of us from the Buddhist side: at first just myself and our two great translators. Now that we have started modern science studies in the monasteries, we have more and more people involved, and this will continue and spread.

"The goals of the dialogue are on two levels. One is the academic level. Generally speaking, our world is highly developed in the knowledge of external matters, though of course there are still many things to explore. But modern science seems not to be very advanced regarding internal experiences, whereas ancient Indian thought, Hinduism and Buddhism, has centuries of interest in the internal experiences of the mind. Many people have carried out what we might call experiments in this field and have had extraordinary and significant experiences as a result of practices based on their knowledge. Therefore, more discussion and joint study between scientists and Buddhist scholars on the academic level could be useful for the expansion of human knowledge, external as well as internal. Our past experience shows us that scientific findings are very, very helpful to get a deeper understanding of things such as cosmology. And it seems that Buddhist explanations give the scientists a new way to look at their own field.

Monks gathering outside the temple.

"On another level, I think the very purpose of our life is survival. For the survival of humanity, happiness and calm are crucial. Unhappy, desperate lives will be shorter. One example is last year's tragedy. It clearly demonstrated that modern technology and human intelligence guided by hatred leads to immense destruction. That is very clear. In order to have a meaningful, happy life, material development is very necessary. But at the same time, there's no point in neglecting our internal development. This is what I call human values. Up to now, our material concerns are highly developed. But our internal values have not matured to the same stage as the material world.

"I think Tibetans have knowledge about internal values, and some valuable experience. But they are materially backward, and some of the cause is lack of scientific knowledge. Some still believe the world is flat and Mount Meru is at its center, as mentioned in some scriptures, although there is no reality to that. As a Buddhist, it is important to know reality. Therefore we should know what modern scientists have actually found through experiment and through measurement—the things they have proved to be reality. I think Buddhists, especially some of our old scholars of Tibetan Buddhism, are perhaps a little orthodox. When I first started to introduce scientific ideas, my older colleagues were skeptical of some of the modern descriptions of the world. Although no one explicitly accused me of holding wrong views, I could see some discomfort on the part of some of the monks.

"So we need more knowledge of scientific findings of these things, but it seems this is complementary to our own approach that emphasizes emotional and human values. We cannot just be one-sided. Spirituality is very important, but without touching material values it is not complete. At the same time, science, technology, and material development can't solve all our problems. We need a combination of material development with the internal development of human values, and not simply or necessarily as religion.

"Already some experiments have been carried out that show some practitioners can achieve a mental attitude of calm, even when facing tragic experiences. The results show such people to be happier. That's useful, and it's cheap. You don't need to go shopping or produce anything in a factory. You have more satisfaction, more inner calm, and that's something good.

"The best way would be to try to make these inner values clear, with the help of scientists. Everybody, whether rich or poor, educated or uneducated, as a human being has the potential for a meaningful life. We must explore that as much as we can. We are not talking about Buddha Dharma or any religious tradition. We're simply trying to make clear the potential of the human mind.

"Through that exploration, it becomes obvious that most disturbances are not external but internal, especially in the field of emotion. The antidote for these agents of disturbance must also come through emotion, not through drugs or an injection. To do that, we need knowledge about the world of the mind, of consciousness, of emotions. Eventually we need to develop a kind of awareness that provides the ways and means to overcome negative, disturbing emotions. So that's our goal, the contribution for emotional human values in the long run.

"So there are two levels. One is simply the expression of our knowledge. The second is the level of emotional human values. Thank you."

Arthur Zajonc, the scientific coordinator, thanked His Holiness for his remarks and asked Adam Engle to say a few words in memory of Francisco Varela. The emotion in Adam's voice was clear. "We could spend days talking about Francisco's qualities and attributes, but I'd like to recall his warm smile, his soft voice, and his brilliant mind, and mostly his compassionate heart and his unswerving dedication to the vision of the Mind and Life Institute and the collaboration between Buddhism and science. Francisco actually lived twenty-four hours a day, seven days a weeks, right at the intersection between science and Buddhism. Science informed his Buddhist practice and Buddhism informed his scientific study. Without his compassionate energy and direction, clearly we would not be here today. The Mind and Life Institute would not have flourished and expanded. And so I'd like us all to dedicate this meeting to Francisco's memory, and to welcome his wife and son as honored guests today."

Arthur Zajonc then opened the meeting with an overview of the themes to be covered, addressing himself to the Dalai Lama: "In this meeting, we bring to you and your Buddhist scholars another chapter, another set of themes and developments that have taken place in Western science. We will hear the compelling story that modern biology tells concerning the origins of life and the development of that life over many millions of years until it can support not only life but also sentience—awareness, consciousness, and ultimately human intelligence. There have been enormous strides in this area. On the technical and scientific side, we have here eminent representatives of that set of disciplines. We'll be discussing this against the background of modern physics, because we understand the origins of life as arising out of matter. What is the relationship between the origins of life and the inert matter that is around us? We look forward, of course, to your contribution to this dialogue. There are very important, even essential questions before us. Some of these are philosophical questions concerning the nature of the transitions between the material world, the world of living, and the world of sentience. Also, with new biotechnologies, and especially since the translation of the human genome, very important and essential moral and ethical questions have arisen. We need to have a way of approaching these questions: ethical insights are essential to work with these technologies in appropriate ways in the next century.

"We're also especially pleased and delighted that His Holiness the Karmapa is here with us, as well as the rest of the Tibetan community. We have often longed for a larger representation from the Tibetan scholars."

Arthur then introduced each of the participants briefly to His Holiness and sketched their roles in the discussions to come. Allow us here to introduce them more formally to readers:

Michel Bitbol is presently Directeur de recherche at the Centre National de la Recherche Scientifique in Paris, France. He is based at the Centre de Recherche en Epistémologie Appliquée (CREA). He teaches the philosophy of modern physics to graduate students at the Université Paris I (Pan-

Michel Bitbol, the philosopher of the conference.

théon-Sorbonne). He was educated at several universities in Paris, where he received successively his M.D. in 1980, his Ph.D. in physics and biophysics in 1985, and his "Habilitation" in philosophy in 1997.

He worked as a research scientist from 1978 to 1990, specializing first in the hydrodynamics of the blood flow in arteries and then in the microstructure of the red blood cell membranes studied by EPR and NMR techniques. In 1990 he turned to the philosophy of physics. He edited texts of general philosophy and of quantum mechanics by Erwin Schrödinger and published a book entitled *Schrödinger's Philosophy of Quantum Mechanics* (Kluwer, 1996).

He also published two books in French on quantum mechanics and on realism in science, in 1996 and 1998 respectively. More recently, he has focused on the relations between the philosophy of quantum mechanics and the philosophy of mind. He published a book on that topic in French in 2000, and worked in close collaboration with Francisco Varela. In 1997 he received an award from the Académie des Sciences Morales et Politiques for his work in the philosophy of quantum mechanics. He is presently learning some Sanskrit in order to get a better understanding of basic texts by Nagarjuna and Candrakirti, for a new philosophical project on the concept of relation in physics and the theory of knowledge.

Steven Chu is the Theodore and Frances Geballe Professor of Physics and Applied Physics at Stanford University. He did his Ph.D. and postdoctoral work at Berkeley before joining AT&T Bell Laboratories in 1978.

While at Bell Laboratories, he did the first laser spectroscopy of positronium, an atom consisting of an electron and a positron. Also at Bell Laboratories, he showed how to cool atoms with laser light (optical molasses) and demonstrated the first optical trap for atoms. Known as "optical tweezers," it is also used to trap microscopic particles in water and is widely used in biology. His group demonstrated the magneto-optic trap, the most commonly used atom trap.

Chu joined the Stanford physics department in 1987. His group at Stanford made the first frequency standard based on an atomic fountain of atoms and developed ultrasensitive atom interferometers. Using the optical tweezers, Chu developed methods to simultaneously visualize and manipulate single biomolecules. His group is also applying methods such as fluorescence microscopy, optical tweezers, and atomic force methods to study the protein and RNA folding and enzyme activity of individual biomolecules. Notable findings include the discovery of "molecular individualism" and the chemical/kinetic basis for "molecular memory."

Chu has received numerous awards, including co-winner of the Nobel Prize in Physics in 1997. He is a member of the National Academy of Sciences, the American Philosophical Society, the American Academy of Arts and Sciences, and the Academia Sinica. He is also a foreign member of the Chinese Academy of Sciences and the Korean Academy of Science and Engineering.

Ursula Goodenough is professor of biology at Washington University in St. Louis, Missouri. She was educated at Radcliffe and Barnard colleges, at Columbia, and at Harvard University, where she received a Ph.D. in 1969. She was assistant and associate professor of biology at Harvard before moving to Washington University. She teaches a cell biology course for undergraduate biology majors and, with a physicist and a geologist, also teaches a course, "The Epic of Evolution," for nonscience students.

Her research focuses on the cell biology and (molecular) genetics of the sexual phase of the life cycle of the unicellular eukaryotic green alga *Chlamydomonas reinhardtii* and, more recently, on the evolution of the genes governing mating-related traits. She wrote three editions of a widely adopted textbook, *Genetics*, and has served in numerous capacities in national biomedical arenas. She joined the Institute on Religion in an Age of Science in 1989 and has served the organization since then in various executive capacities. She has presented and published papers and seminars on science and religion in numerous arenas and wrote a book, *The Sacred Depths of Nature* (Oxford University Press, 1998), that offers religious/spiritual perspectives on nature, particularly biology at a molecular level.

Eric Lander, Ph.D., a geneticist, molecular biologist, and mathematician, is a member of the Whitehead Institute and the founder and director of the Whitehead Institute Center for Genome Research, one of the world's leading genome centers. He is one of the driving forces behind today's revolution in genomics, the study of all of the genes in an organism and how they function together in health and disease. Dr. Lander has been one of the principal leaders of the Human Genome Project. He is also professor of biology at the Massachusetts Institute of Technology. Under Dr. Lander's leadership, the Center for Genome Research has been responsible for developing most of the key tools of modern mammalian genomics. The center led the effort to develop genetic and physical maps of the human and mouse genomes, providing a critical foundation for both genome sequencing and the study of disease genetics, and made the largest contribution to the international project to sequence the human genome, producing about 30 percent of the total sequence.

In addition, the center launched a revolution in the study of genetic variation and its application to understanding human disease and led a collaborative effort to identify more than 1.5 million sites of common genetic variation in human beings. It also organized the ongoing effort to sequence the mouse genome. The center has made these tools immediately and freely available to the scientific community, with the aim of accelerating progress in biomedical research.

Dr. Lander earned his B.A. in mathematics from Princeton University in 1978 and his Ph.D. in mathematics from Oxford University in 1981. In addition to his work in biology, he was an assistant and associate professor of managerial economics at the Harvard Business School from 1981 to 1990. Dr. Lander was named a Rhodes Scholar in 1978 and received a MacArthur Foundation Fellowship in 1987 for his work in genetics. He was elected to the U.S. National Academy of Sciences in 1997, the U.S. Institute of Medicine in 1998, and the American Academy of Arts and Sciences in 1999. He has received numerous awards and honorary degrees, and has served on many advisory boards for governments, academic institutions, scientific societies, and companies.

Pier Luigi Luisi is professor of biology at University of Roma 3. He was previously professor of macromolecular chemistry at the Swiss Federal Institute of Technology, one of the most prestigious technical universities of Europe. Earlier, he traveled and worked in Italy (where he got his degree), the United States, Sweden, and the former Soviet Union. His major research interest is in the phenomena of self-assembly and self-organization

Matthieu Ricard and his mother, Yahne Le Toumelin, who is also a Buddhist nun and a well-respected painter.

of chemical systems, and the emergence of novel functional properties as a consequence of the increase of molecular complexity.

He is presently well known in the field of the origin of life and the origin of protocells, where he combines a hard-core experimental approach with the basic philosophical questions about minimal life. In this field, he is a follower of the theory of autopoiesis as proposed by Varela and Maturana and developed it further into the experimental chemical autopoiesis. Professor Luisi is also responsible for an intense program that bridges science with humanities, the Cortona-Weeks project. He is the author of more than 300 scientific papers and also of literature, including children's books.

Matthieu Ricard is a Buddhist monk at Shechen monastery in Kathmandu and French interpreter since 1989 for His Holiness the Dalai Lama. Born in France in 1946, he received a Ph.D. in cellular genetics at the Institut Pasteur under Nobel Laureate François Jacob. As a hobby, he wrote *Animal Migrations* (Hill and Wang, 1969). He first traveled to the Himalayas in 1967 and has lived there since 1972. For fifteen years he studied with Dilgo Khyentse Rinpoche, one of the most eminent Tibetan teachers of our time.

With his father, the French thinker Jean-François Revel, he is the author of *The Monk and the Philosopher* (Schocken, 1999), and with the astrophysicist Trinh Xuan Thuan, he is the author of *The Quantum and the Lotus* (Crown, 2001). He has translated several books from Tibetan into English and French. As a photographer, he has published several albums, including *The Spirit of Tibet* (Aperture, 2001) and *Buddhist Himalayas* (Abrams, 2002).

Arthur Zajonc is a professor of physics at Amherst College. He has been a visiting professor and research scientist at the École Supérieure in Paris, the Max Planck Institute for Quantum Optics, and the universities of Rochester, Innsbruck, and Hannover. He is a founding member of the Kira Institute, an organization that explores the interface of science, values, and spirituality. He is a consultant with the Fetzer Institute and president of the Anthroposophical Society in America and the Lindisfarne Association.

He is the author of *Catching the Light: The Entwined History of Light and Mind* (Oxford University Press, 1995), coauthor with George Greenstein of *The Quantum Challenge: Modern Research on the Foundations of Quantum Mechanics* (Jones and Bartlett, 1997), and coeditor of *Goethe's Way of Science: A Phenomenology of Nature* (State University of New York Press, 1998).

Tenzin Gyatso, the 14th Dalai Lama, is the leader of Tibetan Buddhism, the head of the Tibetan government-in-exile, and a spiritual leader revered worldwide. He was born on July 6, 1935 to a peasant family in a small village called Taktser in northeastern Tibet. He was recognized at the age of two, in accordance with Tibetan tradition, as the reincarnation of his predecessor, the 13th Dalai Lama. In Tibetan Buddhist tradition, the Dalai Lamas are manifestations of Avalokitesvara, the Buddha of Compassion, who chooses to reincarnate for the purpose of serving human beings. Winner of the Nobel Prize for Peace in 1989, he is universally respected as a spokesman for the compassionate and peaceful resolution of human conflict. He has traveled extensively, speaking on subjects including universal responsibility, love, compassion, and kindness.

Less well known is his intense personal interest in the sciences; he has said that if he were not a monk, he would have liked to be an engineer. As a youth in Lhasa it was he who was called on to fix broken machinery in the Potala Palace, be it a clock or a car. He has a vigorous interest in learning the newest developments in science, and brings to bear both a voice for the humanistic implications of the findings and a high degree of intuitive methodological sophistication.

Thupten Jinpa was trained as a monk at the Shartse College of Ganden Monastic University, South India, where he received the Geshe Lharam degree. In addition, Jinpa holds a B.A. Honors in philosophy and a Ph.D. in religious studies, both from Cambridge University. He taught for five years at Ganden and worked also as a research fellow in Eastern religions at Girton College, Cambridge University.

Jinpa has been a principal English translator to H. H. the Dalai Lama for nearly two decades and has translated and edited numerous books by the Dalai Lama, including *Ethics for the New Millennium, Transforming the Mind, The World of Tibetan Buddhism,* and *The Universe in a Single Atom.* His own publications include works in both Tibetan and English, the most recent being *Self, Reality and Reason in Tibetan Philosophy.*

Jinpa teaches as an adjunct professor at the Faculty of Religious Studies at McGill University, Montreal. He is currently the president of the Institute of Tibetan Classics and heads its project of critical editing, translation, and publication of key classical Tibetan texts, aimed at creating a definitive reference series entitled The Library of Tibetan Classics.

B. Alan Wallace is president of the Santa Barbara Institute for Consciousness Studies. He trained for many years as a monk in Buddhist monasteries in India and Switzerland, and he has taught Buddhist theory and practice worldwide since 1976, also serving as interpreter for numerous Tibetan scholars and contemplatives, including H. H. the Dalai Lama. After graduating summa cum laude from Amherst College, where he studied physics and the philosophy of science, he earned his M.A. and Ph.D. in religious studies at Stanford University. He has edited, translated, authored, and contributed to more than thirty books on Tibetan Buddhism, medicine, language, and culture, and the interface between science and religion.

His published works include *Choosing Reality: A Buddhist View of Physics and the Mind* (Snow Lion, 1996), *The Taboo of Subjectivity: Toward a New Science of Consciousness* (Oxford, 2000), as editor, *Buddhism and Science: Breaking New Ground* (Columbia University Press, 2003), *Balancing the Mind: A Tibetan Buddhist Approach to Refining Attention* (Snow Lion, 2005), *Genuine Happiness: Meditation as the Path to Fulfillment* (Wiley, 2005), *The Attention Revolution: Unlocking the Power of the Focused Mind* (Wisdom, 2006), *Contemplative Science: Where Buddhism and Neuroscience Converge* (Columbia University Press, 2007), and *Hidden Dimensions: The Unification of Physics and Consciousness* (Columbia University Press, 2007).

1 / How Real Are the Elementary Particles?

AS OUTLINED IN ARTHUR Zajonc's introduction, the tenth Mind and Life meeting took us on a long journey, from the simplest constituents of matter far into the complexity of human consciousness. This book tracks that journey as it played out over the course of a week in a packed room at the Dalai Lama's home on the threshold of the Himalayas. How to begin tracing this ambitious, seemingly immense arc?[1] We will start with the statement that launched a presentation by Steven Chu, the Nobel Prize-winning physi-

Steven Chu (left) in the hot seat.

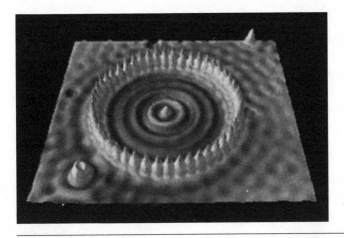

A graphical rendering of iron atoms on a copper surface, as seen throughout a
scanning tunneling microscope.   WITHDRAWN

cist: "The single most important thing we know is that the world is made
of atoms. This is the view that most physicists today, at the beginning of
the twenty-first century, would agree with."

What? some of the monks in the room might have thought. *Are the
mountains, the rivers, the trees, animals and human beings, all made of atoms?
And what about consciousness? Is that too made of atoms?*

The room was filled to capacity, charged with energy, and alight with the
rainbow colors of Tibetan paintings and flowers cut from the Dalai Lama's
garden outside. Steven Chu was looking very relaxed in the presenter's hot
seat, and his boyish ear-to-ear grin signaled eager anticipation as much as
any nervousness about being in the spotlight.

But there was no laughter when Steven Chu showed his slide—a photograph of atoms! At the sight, a murmur passed through the room, and a
small quake rippled through the rows of bald heads. The Dalai Lama's face
lit up and his mouth opened in astonished delight.

"Is this the picture of an atom?" he asked.

"Not a drawn picture," answered Steven, "a real picture! This is a picture of iron atoms placed on the surface of a piece of metal."

"But those are individual atoms?" the Dalai Lama persisted, as if not
quite believing.

"Yes," said Steven. "Individual atoms. Single atoms. Each of these little
bumps is one atom."

"And the wiggles?"

"These wiggles are the electron waves in the metal that are scattering off of these iron atoms. We don't know for sure all the parts of these atoms. But in a certain sense, we can now see them."

"So this is a cluster of atoms?"

"Correct. One, two, three, four, five," answered Steven.

"Do the atoms stay right where they are? Or do they start to stray?" asked the Dalai Lama.

"They would stray if they were at room temperature, because of thermal motions," Steven responded. "You have to get it very, very cold for them to stay still, as in this experiment. However, everything inside the atom is still moving. Within the atom there is a nucleus—the core—and the electrons. The electrons are moving. Within the nucleus there are quarks, and they are moving. The whole frozen atom is just . . ." Steven thought hard and then offered, "It's like a bus that's stopped in traffic. The people inside are still moving, they're hot." He fanned himself for a moment in empathy with these particle passengers,[2] then continued. "So atoms are made of other particles: electrons in the periphery, and protons and neutrons in the core. And even the protons and neutrons are made up of still more simple fundamental particles called quarks. And we know not only that there are six quarks, but each of the six comes in three different colors, as we say. Beyond that, we know of no other particles."

"Wonderful, wonderful!" the Dalai Lama said, smiling with a look of happy complicity. His fascination with atomic particles is notorious, though perhaps a mystery to the monks.

* * *

THE ATOM: A LITTLE HISTORY OF A BIG IDEA

To find the origin of the concept of the atom, we must go back more than two millennia to the town of Abdera in Asia Minor, to the Greek philosopher Democritus (c. 460–370 B.C.). History credits Democritus's teacher, Leucippus of Miletus, with the earliest expression of the idea, but little is known about him other than that he planted a seed that inspired his brilliant student. Democritus started with the assumption of the impossibility of dividing things ad infinitum. Thus the essential components of reality must be particles that cannot be divided further, and those particles were called atoms, which literally means "without division." They cannot be divided further because they do not contain empty space, or "void." In fact, the shape and existence of all things are determined by differences in the shape, position, and arrangement of the atoms and the proportion of void in the substance. It was also believed that the human soul consists of atoms and void.

The philosophy of Democritus was taken up and expanded by Epicurus of Samos (341–271 B.C.) in Greece. Whereas Democritus was highly respected in Abdera, Epicurus was not in the more rigid Greek heartland, and even less was Lucretius (c. 99–55 B.C.), who imported the atomistic theories into Rome a couple of centuries later with his great didactic poem *De Rerum Natura (The Nature of Things)*. Most of the voluminous work of both Epicurus and Lucretius is lost, apparently for the same reason: having been destroyed by their contemporaries and later philosophers and by the government and religious authorities of the time.

What was the reason for this acrimony against them? Interestingly, and as is all too familiar in modern times, it was mostly that their atomistic philosophy did not leave Epicurus and Lucretius with much respect for the gods. Epicureanism was often charged in antiquity with being a godless philosophy because its mechanistic explanations of natural phenomena such as earthquakes and lightning, all based on atoms and their movements, were seen as displacing the will of the gods. Epicurus pointed to the many evils and calamities surrounding us as evidence that the world cannot be under the providence of a loving deity. Even if the gods exist, they have no concern for us, and therefore we should live without regard for them, cultivating tranquility of mind as the supreme achievement of man. While Epicurus was teaching tranquility of mind in ancient Greece, in a very distant part of the world a certain Gautama Buddha was striking a similar chord with an essential point of Buddhism. Chance, synchronicity, convergent mental evolution—take your pick.

For Epicurus, the human mind was also something physical, and he identified mental processes with atomic processes. Whereas atoms are eternals, he reasoned, the mind is not. Just like other compound bodies, the mind ceases to exist when the atoms disperse. This would be so even for the minds of the gods, assuming that they exist.

The Roman Lucretius, who was a great poet, added the dimension of love of nature to the dry philosophy of Epicurus. Particularly in the sixth chapter of *De Rerum Natura*, he articulates the basic atomistic teachings, but also describes some of the controversies between Democritus and Epicurus. Lucretius held that atoms do not move, as Democritus had claimed, in straight lines in all directions and in accordance with the laws of "necessity" (*anangke*). Instead, they move at random and unpredictable moments, and they deviate ever so slightly from a regular course, taking paths and colliding in ways that are not completely determined by necessity but contain some element of chance. This theory of atomic "swerve," or *clinamen*, is a crucial feature of the Epicurean–Lucretian worldview because it provides a physical foundation for the existence of free will. On this basis, Epicurus, Lucretius, and their adherents are remembered as being "able to explain the universe as an ongoing cosmic event—a never-ending binding and unbinding of atoms resulting in the gradual emergence of entire new worlds and the gradual disintegration of old ones. Our world, our bodies, our minds are but atoms in motion. They did not occur because of some purpose or final cause. Nor were they created by some god for our special use and benefit. They simply happened, more or less randomly and entirely naturally, through the effective operation of immutable and eternal physical laws."[3]

This view sounds as if it might have been inspired by Jacques Monod or Stephen Jay Gould speaking from the chair of contingency. And the notion that all is made of atoms appears to be in harmony with Steven Chu's pronouncement that the world is made of atoms. What is new? One difference is that the ancient atomists were philosophers for whom atoms were conceptual entities, the product of logic. Our modern physicists claim to have demonstrated the existence of atoms empirically, through physical experiments. In fact, Steven Chu points proudly at his picture of atoms. We have seen them!

Another advance was the discovery, which Rutherford had made by 1910, that the nucleus accounts for 99.9 percent of the mass of the atom but takes up only a tiny part of its volume. An atom's nucleus occupies the same space as a grain of rice in a football stadium. Void is thus the main reality of an atom. By setting the void inside the atoms, Rutherford parted ways with the Greek atomists, who believed that atoms were small particles of condensed mass with the void outside them.

But this not the end of the story. After Rutherford came Bohr, Heisenberg, and Schrödinger with their quantum mechanics. Their discovery was stunning, and challenged many crucial elements of the age-old picture of atoms and particles. To start with, they realized that atoms and subatomic particles cannot be ascribed any well-defined trajectory. According to Heisenberg's uncertainty principle, if you know the position of a particle with accuracy, you lose any accuracy in the knowledge of the velocity of that particle, and vice versa. But if two or more particles of the same kind (for instance, two electrons) have no well-defined trajectory, we cannot follow their track, and we can easily mistake one of them for another. As a consequence, particles have lost their identity and individuality! Schrödinger was so puzzled by this new truth that he seriously wondered what was left of the concept of an atom or a particle in quantum physics. "What is a particle which has no trajectory or no path?"[4] he asked in dismay.

* * *

"Electrons and quarks are considered our most elementary particles," Steven continued. "We can think of an electron or quark as a tiny little particle that creates a force field. Now this is very strange: our current understanding of these particles is that they have no size. They are infinitely small. The particles are points, and we describe them in terms of their field lines"—in other words, the forces they can exert on other particles in the universe.

The Dalai Lama asked for clarification: "Are you referring to protons, neutrons, and electrons, or to quarks?"

"Protons are not elementary because they are made up of quarks," Steven answered. "Just like the atom is no longer elementary, as it is made up of other pieces."

"And the electron?"

"The electron is elementary, as far as we know. And it is very small. How small is it? If the atom were the size of the Earth, the electron would be smaller than one millimeter. How do we know that electrons are so small? We actually take electrons and throw them at each other. Electrons that have a size would bounce differently from electrons that do not have a size. We can mathematically predict the conclusion that they have no size. They are just points. The particle becomes just the field."

The Dalai Lama was intrigued, but skeptical. "Do they really bounce off of each other? Do they provide an obstruction to an incoming entity?"

"Yes!" Steven said, grinning.

Matthieu Ricard jumped in, unconvinced: "That means they have size!"

"Yes!" Steven repeated, enjoying the paradox.

"This is interesting," the Dalai Lama said. "You speak of something that is only one point, and has no spatial dimension. However, if you hit it from the east side, then it must have a west side. This would imply some type of spatiality."

Steven agreed. "You have to imagine that there's an east side and a west side, a front and a back."

The Dalai Lama pushed this logically: "If it really is dimensionless, then if you hit it from one side, you must simultaneously be hitting it from the other side."

"That's correct," said Steven.

"That's very weird," the Dalai Lama frowned. "There was a fourth-century Buddhist philosopher who refuted the notion of indivisible elementary particles as the simplest building blocks of the physical universe. His argument was that so long as a particle retains a material nature, it will have spatial dimensions and different sides, which are not indivisible. You will probably have to invite this fourth-century Buddhist philosopher to respond to you."

Steven's answer was spontaneous: "Can you be my graduate student?" It was not the first time that an eminent scientist had expressed such wishful appreciation, and the Dalai Lama laughed.

"If I were a bit younger, yes!"

In spite of the humor that graced the discussion, not one of the Tibetan monks seemed comfortable with these strange electrons. Heads shook incredulously, and dismayed faces turned toward the Dalai Lama as if looking to him for some solution to the conundrum. The monks whispered among themselves, and rustlings of consternation passed also through the wider circle of observers. Arthur made an effort to restore silence and attention and Steven tried to move on to the next topic, but the Dalai Lama insisted on pursuing this mystery of the dimensionless particles. He slapped his hand on his fist in a gesture of debate.

"You say that it has not been ascertained that electrons have no spatial dimension. It has simply been ascertained that spatial dimensions have not been possible to detect. The controversy would dissolve if you say that maybe it has a spatial dimension, but too small for us to detect. Is there any reason that this is not an open possibility?"

"That is definitely a possibility," Steven conceded. "This description indeed has fundamental problems. In particular, it leads to a contradiction between our two most cherished theories, quantum mechanics and the theory of gravity."

Particles and the Nature of Reality

Later we returned to this theme. Steven had enjoyed the astonishment that his description of the dimensionless electron caused and was eager to turn the screw further. "We know of three different types of electrons," he said.

"One is the normal electron. If you spend a lot of money, you can make the other two types of electrons by smashing particles together. These other types of electrons are still pointlike, although they have more energy and mass. They can also have either positive or negative charge, particles or antiparticles.

"And now you have a very strange situation: a negative point particle and a positive point particle are attracted to each other. If you bring the particles together on top of each other, they actually collapse and annihilate each other. This is all self-consistent." Steven fell silent. His statement seemed to hang in the air as his audience pondered. Arthur asked if anyone had a question, but no one spoke. Finally, Steven broke the spell with a comment that brought an appreciative laugh from the Dalai Lama: "The fact that we're self-consistent does not mean we're right."

Steven's description of positively and negatively charged particles started us off on a path that would lead eventually to a very deep discussion on the nature of physical properties. But first he explained how the property of an atom's size is related to the charge of its particles. "In an atom, the electron is attracted to the nucleus, which is positive." We were dealing here with normal negative electrons, of course, not the weird and expensive ones that Nobel Prize-winning physicists produce in the laboratory. "The electron wants to get closer and closer to the nucleus as they attract each other. What stops it?" He paused to gauge the effect of his question, but nobody offered a guess.

"Heisenberg's uncertainty principle tells us that, as the electron gets closer and closer to the nucleus, its momentum gets bigger and bigger." To be more precise, it is not the specific value of the momentum that gets bigger, but rather, as you confine an electron to a smaller region, with a small uncertainty in the position, then the *uncertainty* in its momentum increases. "But that momentum requires energy. The nucleus and the electron cannot be packed too close to each other because there is not enough energy."

Steven grinned like the Cheshire cat, and continued, "The energy gain of this momentum has nothing to do with the intrinsic forces of attraction. What's amazing is that the energy gained by the particles getting closer cancels the energy lost in bringing them closer. You lose energy in the process of getting these particles closer, but you gain the energy of the uncertainty principle. The balance of those energies defines the statistical size of the atom, the place where it's mostly stable and the electrons can hardly get any closer." Here Steven added his personal witness to the abstract theory: "I did an experiment on this atom, and the small size that I measured was the same size predicted by quantum mechanics. You cannot get them closer. But now . . ."

Another pause. The afternoon sunlight streamed through the windows, playing over the *tangka* paintings and the statue of the Buddha that watched over the room. The young monks were sitting with mouths open in concentration, but their eyes betrayed a slight sufferance. The Dalai Lama looked happy, and the Karmapa seemed to be following all of this with interest. Still, I wondered how much he could really understand. I told myself that eventually I would need to clarify this, and hopefully I could get an interview with him.

Steven continued, "But now, there *is* a way of getting them closer to each other. This cannot be done with the electron in its natural state, but it can be done by giving the electrons lots of energy in the form of a big momentum. Then, for a brief moment, you can smash them against each other. That's where the ultimate size limit comes: the more energy you give the electron, the smaller the atom can be. So in fact the only limit is how much money the government wants to give you." He punctuated this, like so many of his statements, with a modest shrug, as if to say this was all really very simple.

At this point, Arthur stepped in as the skillful navigator, offering a summary that simultaneously turned the discussion in a new direction. "We've been talking about the size of the atom, which is determined by the balance of the attraction force and the dispersion effect of the Heisenberg uncertainty principle. We can see that size is not an attribute of the elements or of the elementary particles, but we can also understand how size or extension emerges from these particles." He then invited a comment from philosopher Michel Bitbol.

"We have to be very conscious of one point," Michel offered, "namely that the uncertainty relation derives from the very idea that the two experimental devices, one of which can measure the position and the other the momentum, are incompatible with each other. Those so-called properties of the atom are relative to the type of apparatus you use to measure them. Thus, I believe that the uncertainty principle is a manifestation of a kind of interdependence, whose special consequence is the impossibility of disentangling the properties on which the measurement focuses from the apparatus that performs the measurement."

Steven had a different angle of observation from years of working experimentally with these concepts: "The uncertainty principle isn't only something derived, but also a primary thing that we can test by experi-

Michel Bitbol listening to Matthieu Ricard.

ment. We test these complementary properties—position and momentum—all the time in our daily experiments with high-energy accelerators, and we always find that the higher the energy, the higher is the momentum. We see this all over the place. It's hard to escape it. So in my mind this is really a foundational thing that's experimentally testable."

Arthur agreed emphatically, but then persisted in his effort to build a bridge to the Buddhist view. "Michel's point, though, is one we hope bears on Buddhist epistemology: namely, what is the relationship between attributes or properties and the entity that bears those properties?"

As Arthur had hoped, the Dalai Lama responded. "In fact, this mutual dependence between an entity and its attributes is used by the Madhyamaka, or Middle Way, school as one of the principal arguments to negate the intrinsic reality of things." When Arthur asked him to expand on this, the Dalai Lama shrugged modestly and said, "There's not much to say," then laughed at his joke on the vastness of a subject so central to Buddhist philosophy. But he continued, "Things and their properties are mutually dependent. The relationship between them is analogous to the relationship between the whole and its parts. For example, one cannot conceive a whole independently from the parts that constitute the whole. But the notion of parts is again dependent upon the notion of a whole. So there is a kind of a mutual dependence and neither is epistemologically prior."

His Holiness the Dalai Lama commenting on Steven Chu's electron.

Smiles of satisfaction lit the faces of the monks. Finally a familiar concept had surfaced. The Dalai Lama went on, "In the same way, one can speak of an entity only in relation to attributes, and one can speak of attributes only in relation to an entity. Once you have conceptually removed all the attributes, it is nonsensical to speak of what remains."

"It's nonsensical in Buddhist thinking," Arthur qualified diplomatically.

At this point, Alan Wallace stepped out of his role as a translator to offer some reflections in his own voice. Alan speaks so rapidly that it would be a hard challenge to follow him if his speed were not matched by extraordinary clarity. I remember the first time I heard him talk, and my wonder at the agility and force of his argumentation. His hands move at almost the same velocity, dancing to the logic and tracing marvelous arcs and angles in the air.

"Looking at the history of science, in the 1870s there was great confidence among scientists, including Lord Kelvin, in the existence of ether as an all-encompassing, light-bearing medium. When Michelson and Morley's experiment to demonstrate the existence of ether came out negative, the whole notion of ether was jettisoned, and ever since then nobody has taken it seriously. However, for almost a century now we have accepted general relativity and the notion of a medium that warps and woofs, depending on things passing through it." Alan's own hands are warping and woofing as he speaks, an elegantly expressive demonstration of the curvature of time-space. "Nobody generally calls that 'ether,' but one could. They could have resuscitated the word 'ether' as a way of saying that the time-space continuum is not mere nothingness.

"If we go back to the electron, when this term was first coined, our understanding of electrons was very different. It did not include the different types of electrons that Steven has identified, or particles and antiparticles with a different charge. My point here goes back to the earlier question on the relationship between an entity and its properties: how many of its attributes does a phenomenon need to lose before you would say it has become something else? Are all of its attributes necessary? Could it lose one, or two, or three? And is this purely a matter of convention, or does something determine objectively how much can be taken away?"

The reflective pause that followed Alan's questions was filled with a buzz of Tibetan voices. A lively debate had been taking place on the Tibetan side, with the Dalai Lama calling for opinions from the senior monks huddled beside him. Thupten Jinpa stepped back a moment from his role at the center of the philosophical fray to put his interpreter's hat on again and fill us in. "The discussion we've been having here concerns the fact that there are two levels of analysis in Buddhist thought, particularly in the Tibetan tradition," he said. "On one level, the analysis lies within the scope of conventional reality and does not question the ontological status of things or touch upon what Buddhism labels as ultimate reality. Given that we are asking the question at the level of this conventional reality, does it make sense to speak of an entity when you conceptually take away all its properties? There seems to be a divergence of opinion among us."

Steven Chu offered his own view: "In physics when we talk about simple particles, 'simple' means that they only require a few descriptors, such as charge, mass, and spin angular momentum. You can count them on your fingers. As far as we know, those few properties account for everything that the particle does. And so we just step back and say that is the only knowledge we're able to get from those particles. We haven't been able to see anything else with our instruments, and there's no other property that is missing from our explanation. There may be more properties, but we don't see them and we don't need to postulate more because we can explain all our experiments."

Alan's eyes flashed as if he had detected a chink in the armor of Steven's argument, and his hands went on the attack. "But you never measure an electron apart from its properties, right? That being the case, since the properties of an electron are what you actually measure, but you never measure the electron that has all the properties, how seriously do you take the electron itself? Is the 'electron' simply a kind of shorthand for a nest of properties that you can measure? Or do you think that the electron is really something out there that has those properties?"

"We don't actually ask that question!" Steven answered boldly and honestly from an experimentalist's point of view. "The way we tackle that question indirectly is to say that the descriptors we are using—the mass, the charge, the momentum and position—are properties of that electron. Another electron will have a different momentum and different position, but all electrons have the same charge, mass, and spin angular momentum. In other words, all the electrons we see have those characteristics, but they could have different position and momentum. Since position and momentum are different for different electrons, they are not intrinsic properties. The intrinsic properties are charge, mass, and spin. And since those are the only things we know about the electron, and there seems to be nothing missing, then you say: that's it."

Steven threw Alan a look of friendly defiance, but Alan seemed unimpressed. He persisted, "If you had a particle that had two out of three of these properties, but the charge was different, would you call it something else?"

"If the charge was different we'd certainly give it another name," Steven conceded.

And now Alan returned to his core question, as if nothing had passed in the meantime. "Is the electron as real as those three properties that it has?"

"The electron is the sum of all those three properties." Steven stood firm. "And there's one other property, the lepton number," he added, not to let the spirit of debate override scientific accuracy.

It was now time for Matthieu Ricard to come to the aid of the Buddhist front. Whenever he spoke, the Tibetan monks smiled, as if proud of their French brother in maroon robes. Here he offered a traditional Buddhist exposition of how a property is owned:

"One possibility is that things own their properties like a farmer owns a cow. They are independent of the properties, but they own them. The second possibility is that things own their properties like you own your own body. In that case, if your body is the sum of different properties, when you remove the properties, nothing is left. In the case of the electron, the materialist view says that the electron owns its properties like the farmer owns the cow, which means it is still there when you remove all the properties." Matthieu adjusted his robes with a sweeping motion and continued. "The Buddhists say that there are no intrinsically existing entities

Matthieu Ricard helping the Buddhist front.

separate from their properties. All you can do is to describe the different observable phenomena, like the mass, the spin, and so forth. But you cannot describe an intrinsically existing entity that bears those properties."

Michel Bitbol stepped in here: "I think Steven is a very prudent philosopher and physicist. He says, 'When I see those properties, I say it is an electron.' In other words, it is meaningless to speak of an electron independently of its observable properties. However, I wish to remind you that some physicists and philosophers of physics have a much more ontological and much less prudent view of things, especially those who follow Bohm's theory of 1952. They do believe that something really exists behind the properties and somehow apart from them. Unfortunately, this view yields strange consequences. When they were confronted with certain experiments on interferometry done with neutrons—particles that are massive but have no charge—they found that the results could be predicted only if they assumed that the mass of one neutron was dispersed throughout space, although the neutron itself is still supposed to be located precisely. Bohm's interpretation ascribes to the neutron a permanent, precise position that we can measure, and yet its mass is everywhere! According to the supporters of this interpretation, this looked like nonsense. But some of them finally accepted such a consequence, declaring that the neutron must be a 'bare particular,' a particular without properties, although from time to time, the neutron's properties are concentrated and come back to their original bearer. So, as you see, some physicists do believe in the idea of an entity that can survive, so to speak, without its properties!" Accustomed as he is to the wildest intellectual positions of the world of hardcore philosophy, Michel shook his head. "I find this very strange," he confessed. And then he began a very involving, complex series of arguments, which I can only summarize briefly below.

Reality and the Purpose of Science—A Philosopher's View

In fact, Michel Bitbol has long questioned whether the physicists' claim that nature is made of atoms, electrons, quarks—particles construed as building blocks of larger, more complex entities—should be taken literally. Not that he would disregard Steven's photographs of atoms; but he reminded us that these impressive images are only what philosophers call "values of observables"—phenomena arising from the relation between the microscopic environment and the tunneling microscope, a complex apparatus whose construction and interpretation is guided by highly elaborated theories. He asked us to ponder whether the purpose of science is to provide answers about what things are in themselves, or if science is only our most advanced way of developing a pragmatic, conventional knowledge of phenomena sufficient to guide us in our actions.

This question is also important in a Buddhist context. To quote Nagarjuna, "the Dharma is based on two truths: a truth of worldly convention and an ultimate truth." Which of these two types of truth does science offer us? Michel made it clear that he believes the truths of science side with worldly convention, providing us with collective conventions about what to assume and how to behave in order to communicate and act most efficiently. They are not ultimate truths about what the world *is* or *is not* in the absolute sense.

He admitted that few scientists are ready to renounce their belief in the ability of science to reveal something about what the world *is*. However, physicists know that their particles, including atoms, are only fleeting phenomena that emerge in the context of an interaction with a macroscopic apparatus. No individuality can be ascribed to each particlelike phenomenon, and no permanent identity can be ascertained between two experimental phenomena. The disturbing thing for the layman is that physicists still speak in popular science journals of "particles of matter," as if nothing had changed since the time of classical physics. True, when they use the word "particles" in their laboratory, it is with so many qualifications that very little is retained of the familiar notion of material body. But few people outside the specialized circle of physics are aware of this deep change.

With this warning in mind, Michel Bitbol addressed himself to the Buddhist scholars, inviting them not to be too much impressed by the physicists' way of speaking. To talk about particles as little bricks of matter is *only* a way of speaking that is used to allow some connection between physics and everyday forms of thought. This terminology is useful as a glue between these two strata of language and thought, but, Michel affirmed, it has the drawback that it leaves an enduring feeling of crisis in our view of the world due to its bad fit with both the formalisms and the microscopic phenomena.

Like all professional philosophers, Michel Bitbol cannot withstand the temptation to invoke the giants of his field. He reminded us that Immanuel Kant discarded the dogmatic doctrine he had inherited from his teachers, according to which phenomena can only show us things as they appear, whereas our reason can reveal things as they are. Yet Kant still ascribed an important role to concepts: although concepts cannot grasp the intrinsic essence of things, they help us to look for a certain relational pattern between phenomena and to retain only those phenomena that agree with it. This relational pattern is crucial indeed, because as soon as we have picked out phenomena that agree with it, nothing can prevent us from thinking and behaving as if these phenomena were completely detached from us.

To sum up, Kant did not define objects as parts of what there *is* out there, but as configurations of phenomena about which we can elaborate impersonal knowledge. With this in mind, we can revisit the description of atoms or the behavior of electrons that Steven Chu offered. Such descriptions offer knowledge about phenomena that can be shared with any subject, anywhere and at any time. But by no means can they be considered as knowledge about things as they *are*. This view lies somewhere between dogmatism and skepticism, a middle way between the belief that science informs us about being and the opposite but equally extreme view that since science does not inform us about being in itself, then it has nothing objective to say at all.

This Kantian conception offers an interesting parallel with Buddhism. In his reading of Buddhist epistemology, Michel was intrigued and delighted to find the view that the only objects of knowledge to which reality can be attributed are the instantaneous particulars (*sva-lakshana*) that are apprehended by perception or pure sensation. Then, another family of objects of knowledge is derived by inference from the sequence of these instantaneously sensed particulars. These are generalized concepts (*samanyalaksana*), which include substantiality, qualities, relations and spatiotemporal location, criteria of classification, and linguistic determinations. These derived objects of knowledge are considered purely mental constructs, real only in terms of intersubjective consensus. They have no reality apart from our own collective conceptualizing activity.

* * *

THE DHARMA OF ATOMS

The very early Buddhists had a notion of elementary particles rather close to that of the Greek atomists, although the term "atom," as it corresponds to the Western notion, was never used. These elementary particles were understood as the basic constituents of everything in the world. As Matthieu Ricard mentions in his book with Trinh Xuan Thuan, *The Quantum and the Lotus*, the question of atoms was discussed by Buddhist philosophers such as Nagarjuna and Aryadeva in the second century, by Vasubandhu in the fourth century, and Candrakirti in the eighth century. The early Indian Buddhists thought that atoms (we will use the word even if they didn't) were arranged like grains in a cup to compose matter. Matter looks continuous to us, they reasoned, simply because we cannot examine it closely enough—"just like a field looks like a large patch of green from a distance, whereas in fact it consists of a multitude of distinct blades of grass."

Soon enough, however, Buddhist logic developed a rebuttal of the notion of indivisible particles. Matthieu Ricard, who earlier protested the view that electrons do not have dimensions, explained again the logic of his ancient teachers: "Let's suppose that these two indivisible particles come into contact. . . . Then the western side of a particle would first touch the eastern side of a second particle. But if these particles have a west side and an east side, then they are made up of parts and so cannot be indivisible."

These arguments are also expounded in the recent book by His Holiness the Dalai Lama, *The Universe in a Single Atom*,[5] where His Holiness recalls that the notion of atoms was already present in Buddhism in the second century B.C.E. However, an "atom," according to the school of Vaibhashika, is seen as a collection of eight substances, namely the elements earth, water, fire, and air, as well as the so-called derivative substances, form, smell, taste, and tactility. According to Vaibhashika, when such atoms composed of eight substances aggregate to form objects, they do not touch each other.

The ancient Greek view also assumed that there is empty space between atoms—so atoms do not need to come in contact with each other. The Buddhist critique of atomism soon took a quite different avenue. The mechanistic view of atoms shifted very rapidly, particularly with the advent of Mahayana Buddhism, to one in which these units of the material world were replaced by "units of perception." This reflected the general Mahayana preoccupation that the reality of our world is the reality of our experience.

Thus, according to this school, the notion of physical particles disappeared from Buddhist philosophy to be replaced by the term *dharma* (not capitalized to distinguish it from the Dharma, the teaching of the Buddha). "The term *dharma*," writes Alfonso Verdu, "establishes itself as designating the basic, primordial constituents of the conscious stream of individual being, this considered as subject of world-conscious-experience."[6]

These dharmas are thus elements of awareness: a set of dharmas constitutes our present bundle of perceptions. As Verdu states: "The present existence of the *dharmas* is momentary inasmuch as they are manifested only in association with one another—their evanescence is then restricted to their actualized manifestation in bundles or aggregations in the indivisible moment of the present."

As Hirakawa Akira observed,[7] "dharma" itself suggests the quality of truth. However, particularly in early Buddhism, these dharmas were understood to arise through dependent origination and thus as not substantial. Dependent origination refers to the existence of things in a state of mutual interdependence: one's existence is dependent upon and conditioned by others. This dependence is a general rule for all existence, and the foundation of the notion of emptiness. Things are "empty" of inherent existence because nothing exists independently of everything else, and the world consists of elements that are mutually in cooperation with each other.

Two of the fathers of modern quantum mechanics looked in this same direction. Niels Bohr said, "In our description of nature, the purpose is not to disclose the real essence of phenomena but only to track down, so far as possible, relations between the manifold aspects of our experience." And Heisenberg added, "The world thus appears as a complicated tissue of events, in which connections of different kinds alternate or overlap or combine, and thereby determine the texture of the whole."[8]

Not all Buddhist scholars share Verdu's view about the importance of dharma particles in Buddhism, but certainly the notion of particles in Buddhism is an intricate matter. For example, one finds in Kalachakra texts the notion of "space particles." Space, as a supportive element, enables the existence and functioning of the four basic elements (earth, air, fire, water), and space is seen—as His Holiness states in the previously mentioned book—"not as total nothingness . . . but as a medium of 'empty particles' or space particles which are thought of as extremely subtle 'material' particles." And he adds: "According to the Kalachakra cosmology, we can trace the origination of all material objects backwards to the level of the empty particles, which themselves take us back to the origin of the universe. These space particles . . . 'contain' the substantial cause of all the matter that exists within a particular universe system." His Holiness also notes that the term "particle" is perhaps not appropriate, but that there is little help from the text to define such entities further.

Thus, in talking about particles we have unintentionally plunged into the very essence of Buddhist philosophy. In Buddhism, the notion of elementary units cannot be isolated from the general context of philosophy, and it immediately raises questions about reality, perception, consciousness, and the notion of self.

The Special Nature of Physics

After Michel Bitbol's contribution, heads were lowered and faces somber, as so often happens after serious talks by philosophers. I looked at the Dalai Lama's face, always so expressive and transparent, and he also was lost in thought, unsmiling. The face of His Holiness the Karmapa was particularly interesting. His large, deep eyes were slowly scrutinizing the scientists, then moving back to the Dalai Lama as if waiting for his reaction.

This man fascinated me. I had met him a couple of years before at the Gyuto monastery near Dharamsala, an experience I will never forget. The previous Mind and Life conference had taken place shortly after the Karmapa's dramatic arrival from Tibet, and we were invited to an audience with him. We were assembled in a large theater, facing the screen. At that time, the Karmapa was not yet sixteen, and when he appeared, his manner and speech reflected that magical, hermaphroditic age of adolescence. As he walked across the stage, he seemed barely to touch the floor, and he took his seat with a slow, flexing movement. Then he raised his eyes toward us, and the power of his gaze contrasted sharply with the gentleness

A smiling expression of His Holiness the Karmapa.

of his physical appearance. His eyes were gentle too, but they seemed to trespass on you, and his whole face was calm and glowing with a quiet light. The whole scene lasted only a minute or two, and for the first time in my life I said to myself: *If a god walked the earth, he would look like that.* Then I felt embarrassment and wonder at my own thought, but I continued to enjoy the sight of him for the duration of his short speech. It was a plain speech, without any pretense at wise teaching, grounded instead in the common sense of his limited life experience.

Now the Karmapa seemed much older. Two years had transformed him, at least to my subjective perception. He had shed the magic of adolescence and become a strong, very handsome young man of eighteen. His eyes were perhaps less innocent but equally piercing, and his movements were a tiny bit more self-conscious. Knowing that he had no particular scientific training—at least not in Western science—I wondered the whole time what he could understand of our dialogue. And yet, during the whole week his attention never failed one minute.

In his presentation, Steven Chu addressed questions about the meaning of science, and of physics in particular. If we expected that his views would differ from Michel Bitbol's, it was not because the two men would disagree fundamentally, but because a scientist's angle of observation differs from that of a philosopher. The philosopher looks at things from a meta-level high above; the experimental scientist is grounded in the common sense of empirical observations. And this is what the monks seemed to be waiting for: hard facts. Instead . . . well, let us hear it straight from Steven Chu:

"Physics is very different from all other sciences. It's unique in its ability to make quantitative predictions. For example, on the basis of Newton's laws we can calculate very well how airplanes fly. We can calculate very accurately the properties of atoms and lasers and transistors. Also, physics is unique among sciences in its ability to extrapolate its theories.

"At the beginning of my physics schooling, when I was seventeen years old, my physics teacher told me that we would deal with very simple questions through a combination of conjecture and observations. The ideas could be cast into falsifiable theories, theories that could be tested by experiment, and in principle could be proved wrong. These theories, held accountable to continued experimental challenges, would have to be either refined or abandoned. The more important questions, such as the meaning of life, love, death, and taxes, could not be answered by these methods. Indeed, the small set of questions that physics could address might seem trivial compared to humanistic concerns. Despite the modest goals of physics, knowledge gained in this way could not be discarded by changing fashion. Progress became cumulative through the ultimate arbitrator, the experiment."

Steven emphasized that physics starts with experiential observations, but there is an intricate interplay between observation and mathematical logic. To illustrate, he told the story of how our understanding of the Earth's age has evolved. "There are ways to measure the age of the Earth," he began. "For example, dirt settles down in lakes and eventually forms a type of rock. If you could estimate how fast dirt settles to make a sediment, you could also estimate how old the rock is. From this it was suggested that the Earth is between one and a hundred million years old."

Then Steven explained the alternative theory proposed by Lord Kelvin, the most respected physicist of his time, and his idea of measuring the age of the Earth based on the temperature of the surface—and how this famous scientist arrived at an estimate that was very wrong. This was because he did not know about radioactivity, which is a constant source of energy that keeps the planet hot.

Lord Kelvin's calculations were based on mathematics, but the problem was not the mathematics. The point that Steven was trying to make came across very clearly: mathematics per se is always correct. The reason Lord Kelvin and those before him were wrong lies in the fact that they made wrong assumptions. "Never has mathematics been found to be wrong in the last four hundred years," Steven insisted. "It is logic and mathematics that allow us to test the self-consistency of our observations. If a prediction that we have made is false, then either the experiment is wrong or at least one of the assumptions is wrong. In Kelvin's estimate of the Earth's age, the experiment was right, the mathematics was right, but the assumption was wrong. Mathematics is always right."

Mathematics: Discovery or Invention?

"So what is mathematics?" Steven asked. "First of all, mathematics is something that strikes fear in most people." The Dalai Lama laughed and conceded that he himself was included in that majority. "Mathematics starts with counting." Steven grinned as a row of rubber ducks appeared on the audiovisual screen at the end of the room, which raised a laugh from the Western half of the group and patient, blank stares from the monks. When our interpreters, in preparation for the meeting, had briefed us on potential stumbling blocks in cross-cultural communication, they mentioned mathematics as an area that had often proven rocky in the past. I wondered whether these images of bobbing bathtub toys would help or hinder clarity. But I was also struck by the way the dialogue moved so fluidly between the most complex advances in science and the most basic fundamentals. Here we had a Nobel Prize-winning physicist telling us that two ducks plus two ducks make four ducks, and carrying us from there to the mathematics of quantum mechanics in a very brief tour de force. When the Dalai Lama explained some point of Buddhist philosophy to us, did he also embrace such seemingly disparate levels at once?

By counting, adding, and subtracting ducks, Steven introduced negative numbers. Multiplication led us to square roots. The square root of a negative number brought us to imaginary numbers. "They are called imaginary numbers because there's no reality to them. For example, what number is that which, when multiplied by itself, gives -1? There is no such number; it is an imaginary number. And if you add an imaginary number together with a normal number, you get complex numbers. What does this have to do with physics? As it turns out, the mathematics of quantum mechanics needs complex numbers. When he realized that, the inventor of quantum mechanics felt terrible." In the end, though, he had to accept that the use of imaginary numbers in quantum mechanics was unavoidable.

"The mathematics almost becomes real in itself," Steven continued. "Some things, like electromagnetic waves, can't be described with a mechanical model." He drew waves in the air with his hands. "There were many efforts to go around this problem, but in the end, the mathematical equation had no mechanism that one could point to. People then invented a mysterious mechanical matter that would carry the electromagnetic waves, but it was pure invention because we couldn't conceive that the wave could be just an equation. It seemed too crazy.

"So mathematics is very strange," Steven concluded. He gave another example of this craziness: "Newton's laws allow us to calculate how planets go around each other and how apples fall down to the Earth. But Newton was very disturbed by his own law. And he wrote 'that one body may act upon another at a distance, without mediation, without anything in between, is to me so great an absurdity that I believe no man who has a competent faculty of thinking can ever fall into it.'"

Steven's reference to Newton's stubborn resistance to his own discovery reminded me of a similar reaction by Einstein to the quantum mechanical entanglement. According to quantum mechanics, an entangled particle will instantly "feel" what happens to its distant partner, even from halfway across the universe. Einstein despised this idea of a "spooky action at a distance" mostly because it violated relativity's basic tenet that information cannot travel faster than light.

The Dalai Lama, Thupten Jinpa, and Alan Wallace arguing with Steven Chu.

But back to Steven: "Newton imagined a mysterious force that he could describe mathematically, but he couldn't say anything more about it. And so he described this force in terms of the traditional Euclidian geometry we all know. In this type of geometry, if you imagine rulers laid out in space, you can prove that the sum of the angles in any triangle is always 180 degrees. Now we have to jump ahead 300 years to Albert Einstein. He began to think about gravity and came to the conclusion that space without matter is described by Euclid's geometry; but when you add the presence of matter, it actually bends the geometry of space and time. What does that mean? It means that, if you take the imaginary rulers laid out very evenly in space, these rulers begin to stretch, becoming longer or shorter as you get nearer or farther away from this mass. What about time? Time also stretches and shrinks. A clock that is very near a mass ticks at a different rate than a clock that is far away. This stretching or shrinking of time and space is what we mean by the curvature of time and space. When we describe the Earth orbiting around the sun, we used to think there was a force pulling it around the sun. But scientists say now that the Earth actually goes in a straight line, but the line itself curves. This is totally different.

"Einstein was having a tough time. Finally a friend made him aware of the work of Riemann. More than a hundred years earlier, Riemann had realized that geometry doesn't have to be flat. It can be curved. It's very hard to picture curved geometry in more than two dimensions, but if you were to step into a two-dimensional curved space, the sum of the angles of a triangle would not be 180 degrees. When you spread a triangle over a curved surface like the Earth, the sum is more than 180 degrees." Steven's voice rose as he reached the point he had been aiming for.

"Riemann made it up all out of his own head. How did he do this? What kind of mental state was he in? In fact, Riemann conceived the notion of curved space as he was recovering from a nervous breakdown brought on by the pressure of a geometry exam at the hands of his teacher Gauss, one of the greatest mathematicians who ever lived. This was extraordinary, but more extraordinary was the fact that a hundred years later, Einstein confirmed that the universe does work that way.

"How could this be? How could Riemann have invented something like that out of his own mind?" Steven looked around the room as if seeking sympathy for his concern. It was clear that this was a very important issue for him. "Many mathematicians feel that they don't invent things, they discover them. This implies that all this already exists out there waiting to be discovered."

The Dalai Lama offered some philosophical reassurance: "This can't be a purely mental construct, because it does function as a way of symbolizing what's out there."

"Yes, that's correct," Steven said. "But we don't know what many types of mathematics symbolize. And in geometry there are three different symbolizations: flat space, a curved space that we called closed, where the sum of the angles is more than 180 degrees, and another kind of curved space that is called open. There are three choices, but at most there can be one reality, so it is a discovery, but . . ." Steven hesitated over some qualification, but decided not to diverge from his main conclusion: "The physical laws of nature exist. It's amazing that we can discover these laws and it's amazing that the laws of nature seem to have a mathematical formulation. Fundamentally, we don't know why these theories work so well."

The Object of Physics—Material Bodies or Relations?

Steven Chu's last statement, that physical laws of nature exist, is tantamount to saying that the laws of physics have an objective reality. This seems to reflect a view that Alan Wallace has described as "scientific realism," the belief that the theories of physics represent an independent objective reality—that the laws of nature are revealed by the scientific method through mathematical analysis and empirical verification, and they describe physical events that occur independently of the human mind. In contrast, an alternative view of physical science known as "instrumentalism" holds that "theoretical concepts such as fields, energy, electrons, and so on do not correspond to independent, objective entities; they are simply conceptual constructs, or instruments, that physicists find useful in making predictions about measurements. This corresponds to the idea of 'saving appearances': using a theory to account for observed phenomena, making them intelligible without explaining why they are the way they are."[9]

At this point it is worth bending the space-time constraints of the meeting to include a few extra thoughts from our philosopher Michel Bitbol. He had intended to present them but could not for lack of time. Here is what Michel had to say about the objectivity of science and physical reality. Typically, he started with another question:

"Can physics still be considered as the science whose object is matter? Does physics have an object, and if so, is this object identified with matter as a whole? Our objects are bound to be spatial. They are bound to be material bodies because material bodies are defined as the targets of our motor activity, which constitutes our understanding of space. Moreover, they are the targets of anybody's motor activity irrespective of who she is, where she is, and when she acts. They are spatial *and* objective. This brings us back to Kant: the reason our privileged objects of knowledge have to be material bodies, or spatially extended entities, is that the scheme of motor activity that extracts invariants *in* space is identical to the scheme that defines the structure *of* space."

Michel continued, "The notion of material bodies here is remarkably anthropocentric. It is nothing more or else than a pragmatic notion, useful for human purposes in the immediate environment of mankind. It is a convenient way of organizing phenomena for the sake of anticipating them and sharing intersubjective knowledge about them. I think this idea is quite congenial to Buddhism, and especially to Madhyamaka thought. The Madhyamaka school is very eager to dispel any reification of the notions that guide us in the phenomenal world, including the notion of matter. According to the great second-century Indian philosopher Nagarjuna, the Buddha himself rejected the belief in matter.[10] This is a very strong statement, but it is not tantamount to a flat assertion that matter does not exist. It is only a reminder that we should avoid believing in the *inherent* existence of something substantial we call matter, since what exists is, at most, a sequence of interdependent momentary phenomena that we select and organize in matterlike bundles for the sake of practical orientation and intersubjective communication.

"At our human scale, the motor schemes of reversibility of our own bodies, or those implemented in our experimental devices, give content to the idea that there is something permanent or substantial that retains its identity across space-time, is endowed with properties, and can cause events. Thus, saying that something possesses a certain property is a shorthand description of the wide range of relations in which something might possibly be involved."

Michel concluded, "But, as Heisenberg so clearly showed us, at the microscopic scale all these schemes of reversibility that underpin our belief in the existence of permanent spatiotemporal objects called material bodies are missing. Similarly, the constant, repeatable dispositions that underpin our talk of intrinsic properties, including the so-called 'primary' spatiotemporal features, are missing at the microscopic scale. This is enough to conclude that material bodies are no longer the basic objects of physics.

"Although he was thinking in the framework of classical physics, Kant was already aware of this deep-lying connection between our capacities to know and the way we *define* the objects of knowledge. According to him, phenomena are just by-products of a relation between our faculties of knowledge and the so-called 'thing in itself.' Therefore, he said, we can only grasp relations between these relative phenomena. Kant insisted again and again that a material substance is 'a sum of relations only.'[11] What we call a material substance is nothing other than a center of long-range or short-range forces—a center of potential relations with other centers of forces. One can no longer forget, as Kant would have put it, that the objects of our physical knowledge have no intrinsic foundation."

Reflecting later on our dialogue in Dharamsala, Michel concluded: "One of the major aims of our meeting was to unknot and transform this deep relationship among our conception of science, our cultural background, and our most basic existential attitudes. Buddhism can help science and Western civilization by rendering existentially *natural* those epistemological presuppositions that most easily make sense of our present scientific knowledge. And science can help Buddhism by showing that, whereas with little knowledge the substantialist view of the world proved quite easy to maintain (at least at a purely intellectual level), with wider knowledge it appears increasingly clumsy."

A First Step up the Ladder of Complexity

It is now time to leave the world of atoms and subatomic particles and approach the world of matter. This is a first step up the ladder of complexity to the reality of our physical world and life. But how do we get from atoms to matter? This question is appropriate in light of the apparent contradiction arising from the previous discussion. On the one hand it was said that everything is made of atoms; on the other hand, scientists as well as Buddhist scholars questioned whether atoms are real. If atoms are not so real, then the status of matter may also be in doubt.

This question recalls the Zen anecdote where the master asked a student monk to describe the reality of his walking stick. When the student answered, full of passion and excitement, that the stick is not real, it does not possess any intrinsic reality—whack! He received from that same wooden stick a painful blow to the head that left him wondering indeed about reality. I remember also that in certain Buddhist texts the progression from atoms to matter is described by the analogy of a bowl of rice. Many grains of rice together acquire the form of the bowl, or any other form, though each separate grain does not have that form. But how and why do the rice grains stick together? Old Buddhist texts did not mention the notion of attraction forces among particles.

Arthur Zajonc offered a bridge between the concept of pointlike particles and matter as we normally experience it: "Your Holiness, as a child you probably played with magnets and noticed that it's very difficult to bring them too close to each other. Likewise, it's difficult to bring two points together entirely. In fact, it would be physically impossible to bring two electrons completely together because their charges repel each other. The closer you bring them, the more energy it takes. But there are other forces by which the particles attract each other, and it is possible for these forces of attraction and repulsion to balance each other. In this way, a solid material body of point particles extended in space is created by adding one point particle to the next. They do not touch, but they are held through the forces acting on each other. So a table emerges as an appearance, with extension, but the table is not made of things that have extension. In other words, the property of extension itself emerges from extensionless objects."

In this way, Arthur introduced a very important notion into the dialogue, a cardinal point in the modern science of complexity that would be discussed at length later on: the idea of emergence and emergent properties. Emergent properties are those novel properties that appear when small elements come together to form a higher level of complexity—novel in the sense that they are not present in the constituting elements. In this case, form and macroscopic physical properties would be considered emergent properties of the assembly of small particles.

"So this is how we get extended objects from nonextended objects," Arthur continued. "This property of attraction or repulsion, which we call charge, is very important. But an object may have other properties as well, which can lead to other kinds of appearances. We can then perform an analysis: Are there intrinsic properties that a thing may have? How do we think about those intrinsic properties?"

Again I had to admire Arthur's skill at compromise, remaining faithful to the science even as he depicted physical reality through language and concepts that were accessible to our Buddhist colleagues. (I was reminded of his beautiful book, *Catching the Light,* where he also balanced modern physics intelligently with a more spiritual view.) The question of intrinsic properties is a critical one in Buddhist philosophy, which Matthieu Ricard offered to elucidate.

Buddhism on Particles and Emptiness

Matthieu first reminded us of the atomistic theories of the early Buddhist scholars, in particular their refutation of the notion of indivisible particles, and what light this might shed on the question of the aggregation of matter. "This morning His Holiness asked a very pointed question: If atoms are really dimensionless, with no directions or sides, why should an atom go in one direction rather than another when they come in contact? The question was based, as he mentioned, on the reasoning of Vasubandhu, one of the great luminaries of Buddhist philosophy in the fourth century." Matthieu recapitulated Vasubandhu's reasoning in refuting the atomistic theory: "To build reality we need some kind of aggregation. At least two things have to come together. So the question is: Do they come in contact or not? If they do, how? If two things come in contact, then each must have a front, or an east side that meets a west side. If that is so, if you can recognize some kind of direction, then you don't have an indivisible particle because you can distinguish one side from the other.

"If, instead, particles are like dimensionless mathematical points, so that when two come in contact, all parts of both particles come in contact simultaneously. . . . In that case, what happens?" Here Matthieu grinned wickedly in anticipation of the approaching conundrum. "They fuse. They become one thing because they have no dimension. But if two particles fuse with each other, so do three particles, and then a mountain, and the whole universe could fuse into one particle.

"Finally, you could say that the particles don't need to come in contact to build matter. There can be something in between them. But again the same reasoning applies: if the particles have no dimension and if something can be in between them, then one, two, or three particles can be in between, and also a mountain; the whole universe could fit between these two. So the conclusion is that you cannot possibly imagine a process of aggregation of matter by using mathematical, pointlike particles that have no dimension."

Here it seemed to me that Matthieu threw Steven Chu a defiant look, but if so, it passed in a blink. He turned again toward the Dalai Lama and Arthur, and continued, "Therefore, Vasubandhu concluded that indivisible, concrete particles that have intrinsic, permanent existence cannot account for the physical world. He negated the possible existence of such particles and concluded that the phenomenal world is mere appearances. His reasoning was that, since the microcosmos of these minute particles has no reality, then the notion of aggregation of atomic particles is also devoid of intrinsic reality. Once we have killed an idea, we don't need to kill it twice."

A glorious smile stamped on the face of each of the monks registered a Buddhist consensus, but some of us were a little puzzled by that conclusion. Did Matthieu, who once upon a time was a young Western scientist in our own ranks, really mean to cast doubts on the reality of our physical world? As if Matthieu felt this question in the air, he then qualified his statement: "This notion of being devoid of intrinsic qualities has a parallel, in my mind, with what quantum physics tells us: a particle can be an object that has a location at one point in one set of circumstances, like a stone. At the same time, we can consider it as a wave, like the ripples that a stone makes in a lake. Those properties are quite opposite. The wave is everywhere at the same time, and the stone or particle is in one area. This shows that you cannot say that one attribute or the other is absolutely intrinsic to that phenomenon. It cannot be both localized and everywhere; it cannot be both discrete and diffused. Any aspect or property of a particle that we can analyze—its spin, its mass, or anything else—is an attribute that we designate. If a property were to be intrinsic, then it could not change or be impermanent."

From there, Matthieu went on to condense the Buddhist scenario of reality in a nutshell. "That's how we come to the idea that nothing can be truly intrinsic, totally permanent, autonomous, and independent of anything else. All properties, all observable phenomena, appear in relationship with each other and dependent on each other. This view of interdependence—one thing arising in dependence on another, and their relationship—actually defines what appears to us as objects. So relations and interdependence are the basic fabric of reality. We participate in that interdependence with our consciousness; we crystallize some aspect of it that appears to us as objects. That is a simple way of describing basic reality from a Buddhist perspective, beginning with the refutation of the existence of truly independent, permanent elements of reality."

Intrinsic Qualities

Alan Wallace elaborated on this with a Buddhist analysis of intrinsic properties. "There is a type of analysis," he said, "where one tries to seek out the identity of a thing—a particle, a field, or whatever—by looking at the relationship between the whole and its parts. When one applies this type of analysis to find a phenomenon that is presumed to exist intrinsically, by its own nature, absolutely and objectively, in and of itself—one doesn't find it.

"One possible conclusion is that, since we cannot identify anything absolutely, then everything we see is simply a projection of the mind, simply appearances. According to this school of thought, which understandably

His Holiness and Thupten Jinpa, with Alan Wallace. Often His Holiness's remarks were quite humorous—laughter was not rare in the conference hall.

enough is called the Mind-Only School, the mind is the only true reality. But what if we now turn that same analytical sword back upon the subjective mind, and ask: Where is this mind found? We find many moments of the mind, many mental processes. Which one exactly is 'the mind'? When you start analyzing the mind's parts in search of the whole, you realize that the mind is not to be found either. At this point, you've wiped out the whole objective world, and the whole subjective world as well. You have nothing; you become a nihilist. Or, you come up with another idea."

Like a good storyteller, he paused for a moment of intellectual suspense. His hands stopped too. Then they start again, drawing circles and spirals in the air. "There is another possible way of interpreting exactly the same analysis. Even if there is nothing to be found ultimately or absolutely, in the meantime the entities in question are doing things. For example, photons can make people sick with radiation illness. It's silly to say that those little particles don't exist even though they are making people ill.

"But there's a happy compromise, a middle way. To take one example, physics is a very powerful, internally consistent, and useful description of the world. Moreover, there is empirical evidence for these theoretical entities like the field, particles, and quarks. We have rigorous reason, with very sharp experiments, to believe they exist. Let's not worry about whether they have some intrinsic identity in and of themselves, but let us consider the question: Do they do things? Can you do things to them? Do they do things to other things? If they do, then we can say that these things are real on that level, with respect to the conceptual framework in which we understand them.

"Also, as I'm thinking about these things, my thoughts are as real as anything else. So we can bring the mind back in. We can say: Yes, we have feelings and desires and ideas that exist relationally. The objects that we are investigating also exist relationally. We all have a sense of what an electron is conventionally accepted to be. We've agreed upon that, as we've agreed what a bottle is, or a table. So within this context of conventional agreement, in a coherent theoretical framework, electrons and minds and other things do exist. We attribute their existence conventionally or relatively, without attributing to them any type of inherent, absolute, intrinsic, independent existence. Buddhism settles for that as a happy medium. At least we are happy."

Silence. There was indeed an expression of satisfaction on all the monks' faces, reflecting the Dalai Lama's own look of satisfaction. I glanced at the Karmapa's beautiful face; he seemed happy that the others were happy. Even Arthur looked more relaxed now. The middle path suggested here ran close to his own view. In fact, having accepted this compromise, we can bridge the gap between the whimsicality of particles and the reality of the Zen master's stick, though I wonder whether that monk who received the blow to his head would also say, "At least we are happy."

Reflections

Originally I had thought to limit this first chapter to a discussion of atoms and other elementary particles. Instead, a much broader picture emerged, extending its frame to encompass the meaning of physics—and, by inference, the meaning of science in general—as well as the basic tenets of Buddhism and Western philosophy. Why did the topic expand?

In part, it was because of the caliber of the personalities involved in this dialogue. When you activate people like the Dalai Lama, Steven Chu, Michel Bitbol, Alan Wallace, and Matthieu Ricard, you cannot expect a focused minuet on a single theme, but rather a symphony that resonates to the horizons of the human enterprise. There was, as we have seen, another reason embedded in the topic itself. You cannot talk about "particles" in Buddhism without considering the whole picture of the human mind, perception, and consciousness. Thus, we were obliged to lay down in the first chapter some of the themes that will come up later in more detail.

People of Mind and Life: Daniel Goleman and his wife Tara. He is the well-known author of *Emotional Intelligence*, and also the author of a preceding Mind and Life book, *Destructive Emotions*. Tara Goleman is the author of *Emotional Alchemy*.

Aside from this very general consideration, a couple of other points come to mind.

The atomistic view of the ancient Greeks—with its emphasis on void and movement, and the statement that atoms are the constituents of mind—is in certain respects close to the Buddhist view. All the major atomistic Greek philosophers developed theories about perception that were consistent with their atomistic theories. However, the Greek view drove a wedge between mind and body that led to the schism advocated by Aristotle and all those after him. Meanwhile, Buddhism gradually abandoned the question of units of matter and focused on perception. One might question which of the two approaches was more successful, but that depends on how you define success. As my friend Richard Baker Roshi says, Buddhism could never have accomplished nuclear fission. But then, Buddhism defines success as progress toward enlightenment—the development of compassion, the cessation of suffering, the realization of interdependence. In any case, a view of the world that always keeps the mind in the picture is perhaps more complete than the worldview of Western civilization after Aristotle.

We had raised the question whether modern "atomists" like Steven Chu or Richard Feynman have really said anything conceptually new with respect to the Greek atomists. Our first answer was that modern science is working with hard facts and objective reality. However, this simplistic picture disappears in the fog of the ephemeral character of the atom and the electron, not to mention quarks and the other ethereal particles. Modern science is rich and filled with "irrational" notions that cannot be grasped by our common sense: spaces of many dimensions, photons that are entangled with each other even when miles apart, things that can be particles or waves depending on how you perform the experiment, time and space that are curved and stretchable. . . . The important point made by Steven Chu, that mathematics holds the power of natural laws even before the discovery of such laws, adds a disconcerting element to the picture. Has modern science—or at least physics—made a full turn back to pure philosophy? No, because the experiments are there to support such irrational views. Experiments that confirm the irrational—is this the definition of modern science? Probably yes, in the sense that this is how modern science has to operate in order to link the phenomenology of the world with a satisfying description of reality.

We have also given some sketchy indications of how Buddhism tries to link experiential observation with reality. In the following chapters, we will proceed further in this direction. In particular, we will try to monitor how modern science sees and interprets the complexity of nature. We'll turn our attention away from atoms and consider real matter, with complex molecules and properties, including the property of life. But before doing so, I would like to return to Steven Chu's statement that the world is made of atoms. Steven quoted a famous statement by Richard Feynman in this regard: "everything living things can do can be understood in terms of the jigglings and wigglings of atoms."

Steven later clarified the difference between his statement and Feynman's. To say that our world is *made* of atoms and their movements is one thing; to say that the world can be *understood* in terms of atoms and their dynamics—that is quite another thing. This difference is a fundamental point of this whole book, and the next chapter, which focuses on complexity, emergence, downward causality, and the relationship between structure and properties, is a first step toward the clarification of this difference.

2 / The Emergence of Complexity
AND AN INTERVIEW WITH MATTHIEU RICARD

THE PREVIOUS CHAPTER DESCRIBED how complex matter can be formed from the aggregation of atoms. The aggregation of small particles into larger ones is the basic mechanism by which our life and our universe unfold: cosmic dust particles form stars and planets, atoms build molecules, molecules build complex organic structures that in turn build cells, which aggregate to build multicellular organisms, tissues, organs, mammals. . . . The scientific scenario for the origin of life traces how small molecules, through a continuous accretion of complexity, become entities that can reproduce themselves: the first units of life.

There are two main concepts related to this phenomenon of increasing complexity. One is self-organization; the other is the accompanying feature of emergence. Self-organization involves parts coming together to build an aggregate of larger complexity, a process that is often spontaneous. Emergence has to do with the arising of novel properties—novel in the sense that such emergent properties are not present in the components—when parts come together to build a higher degree of complexity.

At first sight, the idea of a spontaneous increase of order and complexity may appear surprising. We usually think of spontaneous natural processes as producing greater disorder, moving toward greater entropy. But there are processes that spontaneously produce complex structures that are more stable than the disordered mixture of components from which they start.

Notice that we are talking not just about organization, but *self*-organization: a process dictated by the "internal rules" of the system. When a structure is organized by externally imposed forces—as in the construction of a

TV set by an engineer, for example, or the planning of a city by architects—we cannot call it self-organization. Many true examples of self-
organization appear in biology as well as in chemistry and physics. Crystals spontaneously assemble in geometrical forms out of a saturated salt
solution. Surfactant molecules assemble themselves into micelles and
vesicles, a process that creates not only soap bubbles but also the biological
membranes essential to life. Likewise, the formation of a DNA double helix is a self-organization process, where the sequence of the nucleotides of
the two strands determines the rule for self-assembly. Protein folding is
yet another instance of self-organization that is vital to life, as we will see
later in the contribution from Ursula Goodenough. For now, our coordinator Arthur Zajonc guides us into the realm of complexity, having jumped
in because of the last-minute cancellation of Stuart Kauffman, the well-
known theoretician of complexity from the Santa Fe Institute.

The Notion of Complexity

Arthur was quietly concentrating in anticipation of the Dalai Lama's arrival for the next session, but just a few minutes earlier he had been involved in an animated conversation with the biologist Eric Lander. The two
conferred with intense gestures under the curious eyes of the Buddhist
monks, who were trying to decipher the scientists' body language.

The mystery unfolded when Eric moved swiftly toward the first row of
participants and began to organize them. It was a dynamic, if somewhat
confused effort, and it gradually became apparent that his intention was to
simulate a self-organizing pattern for a demonstration that would take
place during Arthur's lecture. Eric imparted certain rules to his victims:
people were numbered, and odd numbers were to stand while even numbers sat; then the participants were to observe their neighbor and stand or
sit depending on the other's position. Eric looked like an orchestra conductor losing patience with his musicians: "Down, down . . . no, you should
stand up. . . . Yes . . . you down, please! Why are you standing up?" The exercise brought hearty laughter from all of us, including the monks, and I
imagined the deities looking down from the *tanka* paintings a little discon-
certed by this unprecedented display. Eventually Eric reached a kind of
peace with his troupe, though it was not clear whether they had finally understood what to do or he had compromised on a simplified score.

The rehearsal concluded hastily when His Holiness entered. Now, attention was focused on Arthur sitting calmly in his chair. He began, "Yesterday we discovered that physics provides a very particular understanding

Arthur Zajonc explaining complexity and self-organization.

of the material world around us, in terms of elementary particles that themselves have no extension, no size, but that can, through forces, create matter. We also discovered the unreasonable or extraordinary success that one can have with mathematics. Mathematics provides a very powerful tool for understanding the forms of nature. Both from the side of physics and from Buddhism, we went very far in trying to catch the ultimate nature of matter, all the way down to the fundamental properties and attributes that are part of the elementary constituents of our world. We saw the difficulties in sustaining the view that these elements are absolutely present and independent of the observer. There was a certain sympathy between the pragmatic approach of modern physics and the philosophical approach of Tibetan Buddhism.

"Today we're going to turn from the nature of matter to the transition to more and more complex systems with many parts. In such systems, new types of phenomena appear. Some of these phenomena are very beautiful." Arthur illustrated his point with a striking photograph of microscopic crystal structures.

The Beauty of Forms

"Sometimes a system is in motion—a flowing river, for example—and the kinds of forms that arise in such dynamic systems are also very beautiful." Here he showed a photograph of the graceful patterns formed by dye dispersing in moving water, then another, equally beautiful, of dramatic cloud systems over the Pacific. He continued with more images: "Not only do we have forms that are characteristic of certain kinds of physical systems, but we see them everywhere in the world of life, in this beautiful image of a bighorn sheep, a glorious goat, or a nautilus shell. In this seashell that has been cut in half, one sees the beauty of a mathematical form. As a scientist, when one sees these forms in the physical world or in the world of life, one immediately is drawn into the research: is there a way to explain these forms, to discover the principles that underlie them?

"Nature does not always provide us with beautiful forms. There are many situations where there is just rubble, a kind of chaos. In a landscape there are many beautiful forms, but if one looks closer, there is perhaps just rubble at the bottom of a mountain." To make his case he showed another perspective on the crystal structure shown first, a more distant view that included copper chloride crystals growing in long pencil shapes, jumbled like sticks thrown on the ground. "If, however, I add a very small amount of an impurity—quite a long polymer molecule—then the structure changes and beautiful forms start to appear. In other words, by small modifications, we can move from a domain where the phenomena themselves are chaotic to a situation where very beautiful forms appear. How do we understand this arena of phenomena? It's about this that we want to speak: the arising of form out of chaos through a particular kind of complexity. Not all complexity gives rise to form, but when certain rules are present, then we are able to see complex systems.

"This is, in some ways, a philosophical point. Historically, Western science has for many centuries attempted to explain large things in terms of small things. In this reductionism, the attributes of the large are seen as somehow identical with the attributes of the small. But our presentations here emphasize how new properties can emerge through a kind of complexity and a set of relationships."

Arthur continued, introducing his main point. "One approach to the emergence of form from complexity that many scientists have used is called self-organization. How much form can arise through complex self-organizing systems? That's what my talk is going to be about. Both emergence and self-organization must certainly play a role in the world of the living. Self-organization, however, is also important for the nonliving world of physics, which is more my world."

Cellular Automata?

Arthur now dove into the substance of his presentation. "In talking about self-organization, there are a number of things to consider. The first is that the emphasis is on the interaction of the parts. Instead of looking only at the parts and thinking of them as primary entities, we focus on the interaction among them and the integration of those different pieces working together. Through that integration, form can arise through the operation of simple rules. You may have noticed this in the example of flying geese. How do they maintain that beautiful formation? We can model calculations to generate the kind of flocking behavior that we see in birds from simple rules.

"I'm going to use a different example, however, one that goes back to the very first studies on the generation and replication of form. One important aspect of life is how it reproduces. John von Neumann and Stanislaw Ulam, two mathematicians—not biologists—were very interested in this question. In 1953 they developed a mathematical description of reproduction, which we are going to demonstrate for you here with this row of people. We practiced before you came a few times to see if we could get these people to act like cellular automata."

The participants seemed nervous about their impending performance as His Holiness looked them over and asked, "Is each one a cell or a gene?"

"Each one is a cell," Arthur explained. "They are abstract, mathematical cells. There are ten identical cells, and they're going to act in a certain way automatically according to a simple rule." Arthur stood up and Eric approached from the other side. A rustle of anticipation swept the room. Would it work? The monks all had big smiles stamped on their faces. Having watched the chaos of our rehearsal, they were no doubt wondering if we could deliver the promised order.

Arthur explained the rules, noting first that it required a certain level of *samadhi,* or concentration, to carry them out: "Everybody in the row is going to look to the left and to the right. If no one is standing on either side, sit down. If one person is standing next to you, stand up or remain standing. If two people are standing next to you, sit down." The game also required an initial condition to get off the ground: two people in the row were chosen to stand while the rest remained seated.

Eric assumed his conductor's stance, raising an imaginary baton. "We're going to do this with a rhythm: look left, look right, then move. Look left, look right, then move." He interrupted his rhythm for a final command: "Move doesn't necessarily mean change! It means do the appropriate move. Even people who are sitting should look left and right. Everybody's following exactly the same rules."

Goldie Hawn, in one of the rows of automata, called out, "And our motivation is . . . ?" It was an actor's joke, but Eric trumped it with a Buddhist answer: "Your motivation is the benefit of all sentient beings!"

When the laughter subsided, Eric continued, "Look left, look right, now move; look left, look right, now move." He repeated the rhythm, which was now punctuated by more laughter as half the participants looked in the wrong direction. It was clear that the emergence of order from chaos is not easily generated in the presence of human intelligence. Eric concluded the performance and returned amid applause to his seat, mumbling that they should have asked the monks to perform the exercise, as they probably had better concentration. His Holiness sat patiently, his eyebrows raised even higher than usual; he was amused, but waiting for an explanation.

Local and Collective Rules

Arthur salvaged the demonstration: "There are number of things that you can actually see in this. First of all, there is a local rule: each person is simply doing the same thing. They don't look far; they don't need to see the end of the line. They just look to their left, to their right, and then act. If they do this properly, an overall form arises, a particular structure: certain ones go up and certain ones go down. Then others go up and go down. A pattern develops through time, and this is a very important factor.

"Another interesting thing is the role that Eric played. If they were completely autonomous, acting completely on their own, they wouldn't really need Eric. He would just educate them, as it were, and then they would act according to their own principle. They would be self-organized—not organized by Eric, but organized according to their own individual rules. This is the basic idea of self-organization: that form can arise from a simple rule."

Arthur then projected a graph with hundreds of cells arranged in rows and columns. Certain cells were marked in black: just two at the top—our initial condition—and then gradually many more, creating triangular patterns running down the grid. If the performance had worked—on a rather larger scale, with a few hundred people in line—and if the ups and downs at each beat of the rhythm had been recorded, this was the pattern that should have emerged.

Arthur elaborated, "If you have different rules or different starting conditions, you can get many different forms, all developing simply on the basis of a local rule." He paused; there was an important qualification. "Not all rules lead to order. This is important. Certain rules and conditions produce patterns; others produce just chaos. Some of the most interesting forms appear at what we call the edge of chaos, between a fully ordered situation that is very regular and one where there is complete chaos. At the edge of chaos we often find emergent patterns that are extremely sensitive to initial conditions. If anything is a little different, just a small change, something very new will happen. This is a characteristic of life: there is a very delicate balance between rigid order and total chaos. Between these two extremes there's a wonderful area of form, evolution, and sensitivity."

Arthur then used the rule-based logic that our troupe had just failed to display in their demonstration and extended it. With a rule that was only slightly more complex mathematically—action based on the state of eight neighbors rather than just two—he showed how copper chloride crystals form in a supersaturated solution.

Self-Organization in Dynamic Systems

"Morphology is the study of these forms, from complex systems in the physical world all the way to organic forms in the world of living things. The person who coined the term 'morphology' was a German poet, and a very favorite poet of mine, Goethe, who lived around 1800," continued Arthur. His way of studying these forms was phenomenological and even contemplative. He didn't know mathematics, but living in this world of shapes—plant, animal, crystal shapes—he wondered what their principles are. And he lived in this study very deeply, almost meditatively, hoping to come to what might in Buddhism be called a direct perception of those principles. His particular approach is a wonderful bridge between poetry, or the humanities, and the sciences. But we're talking today about the way normal science deals with these forms, which is to look for the actual physical processes that generate these rules.

"I'm going to start with a chemical model that was developed by Alan Turing, another very famous mathematician, about the same time that von Neumann was inventing the cellular automata. Turing wrote a paper called 'The Chemical Basis of Morphogenesis,' about the changing of forms and the origins of forms. He starts with a very interesting situation. If you put two chemicals in a liquid and stir them, normally the liquid just sits there and stays the same color, as if you had stirred milk into coffee. But in certain cases, if you pick the right chemicals, the liquid begins to self-organize and generate its own structure and form, and that form will change over time. You can mix it up again, leave it, and it will spontaneously start to generate new forms."

A hint of excitement edged into Arthur's calm voice as he explained. "The two chemicals need to react with each other, and you need two competing reactions. But if that's the only thing that happens, nothing interesting takes place. These chemicals also have to diffuse: they have to move through the medium and become less and less concentrated." Here Arthur illustrated his point with a graphic analogy: a computer model of lions moving across the savannah, where population growth and dispersal over an expanse of land mimicked the chemical reactions and diffusion. The result was a striped pattern of distribution—remarkably similar to stripes found in the natural coloration of certain shells, fish, and butterflies. Presumably in such cases a similar reaction and diffusion are triggered genetically.

Arthur's sequence of images took us back again to chemical reactions in liquids. "You see the spiral that originates here and goes out farther and farther," he pointed out. "This is the so-called Belousov–Zhabotinsky reaction. These forms are produced spontaneously, by chemicals simply coming together, reacting, and diffusing—spirals generated as if out of nothing, from a homogeneous mixture. So not only can this process create static forms, like on a shell, but also it can create moving, changing, evolv-

Some patterns of the Belousov–Zhabotinsky reaction. It is a periodical, oscillating reaction, with patterns forming spontaneously and reproducibly when a complex cocktail of chemicals is mixed (including potassium bromate, sulphuric acid, cerium ions, and several others).

Bénard convections: the artificial beehive structure that is formed spontaneously by heating a layer of silicon oil between two glass plates.

ing forms. This is a very important part of life, of course. We don't only have static, immobile, fixed forms; a kind of evolution and dynamic process takes place as well. This is all self-organized. There is nothing coming from the outside.

"When we thought of order in the past, we thought of it as a kind of perfect order: the motion of the stars and the planets, the motions of machines and so forth, often described by an old type of mathematics. But now we are looking at new phenomena at the edge of chaos, the boundary between chaos and order. Certain features are important at that edge."

* * *

DYNAMIC SYSTEMS AND COMPLEXITY

For many scientists working on the physics of dynamic systems, as well as on artificial life, self-organization and emergence are seen in relation to dynamic systems that change over time. This is the realm of chaos and nonlinear dynamics, self-organized criticality, and self-organization in nonequilibrium systems.[1]

Dynamic systems are generally out of equilibrium, and it seems counterintuitive that a system out of equilibrium may form self-organized structures. A classic example is the formation of "beehive" structures in a liquid silicon layer that is heated between two glass surfaces. The resulting convection creates extended hexagonal structures instead of a more and

Monks debating in front of the temple. The monk asking questions is standing and clapping his hands when making a point, and the monk sitting down is answering the questions. As the standing monk makes a point, he claps his right hand on the left and draw the *mala* (rosary) from the right hand to the left arm, symbolizing drawing all sentient beings out of *samsara*. The standing monk wins if he can demonstrate internal inconsistencies in his opponent's position; the sitting monk wins if he can defend his position consistently with the authoritative texts.

more disordered molecular mixing, as would be expected. The beautiful structure remains visible to the naked eye as long as the temperature difference is maintained.

The temperature at which the hexagonal cells form is a bifurcation point, at which the system may form one pattern or another. This is an important notion in the theory of complexity, introduced by Prigogine and his school.[2] At a critical distance from equilibrium, the system must "choose" between two possible pathways that are equally probable. One pathway leads to an organized structure; the other leads to a region of instability. This notion of the bifurcation point, related to instability and fluctuations, lies at the heart of this branch of science. The emergent organized structures can take the form of oscillations, as in the Belousov-Zabotinski (B-Z) reaction, first observed in the 1950s by the Russian chemist Boris Belousov.[3]

Essential in this field is the work of Ilya Prigogine, who showed theoretically, and in perfect agreement with classic thermodynamics, that order can originate from disorder through self-organization due to oscillations in a system out of equilibrium. Prigogine and his school also developed

Mandelbrot's famous fractals (left), and the graphical rendering of one of their mathematical developments.

the related notion of a dissipative structure—an open system that is itself far from equilibrium, but nevertheless maintains a form of stability. In a pendulum, for instance, dissipation is caused by friction that decreases the speed of the pendulum and eventually brings it to a standstill. But if a flow of energy is continuously provided to the system, the constant structure is maintained.

An important observation coming from these studies is that order and self-organization can indeed originate by themselves: complexity can occur spontaneously. This is why we see matter acquiring new properties when estranged from equilibrium, when fluctuations and instabilities are dominant.[4]

* * *

The Notion of Self-Similarity

Arthur Zajonc continued his presentation: "I've talked about the sensitive dependence of morphogenesis on initial conditions. Another interesting concept that shows up often in nature is self-similarity. A new kind of mathematics was developed to handle this. One of the many contributors to this new mathematics was a scientist named Mandelbrot." Here Arthur showed a series of fractal images derived from Mandelbrot sets. "If I magnify some region of this picture, it looks similar to the overall pattern." He zoomed in to demonstrate.

"There are an infinite number of such pictures that can be generated by this new mathematics. They are often very beautiful and very delicate structures. They are generated simply by a process like that of the cellular automata: a mathematical rule is followed again and again and again. Some of them even look like natural objects. Sometimes it's just chance; other times there's a good reason they bear a similarity."

The Dalai Lama interrupted, "Does this ever actually manifest in nature?"

"Excellent question. This particular form does not manifest, but the property of self-similarity does manifest. It's a very important property in nature. It manifests in many, many places." Arthur proceeded to show sections of the image at increasing magnifications, with similar patterns reappearing at each deeper level. "We could do this for the next hundred years, and you would always be able to find more structure. And it would always look similar. Not the same, but similar."

The Dalai Lama seemed perplexed, and asked Arthur to explain the mathematics underlying the images. He obliged: a complex imaginary number—one containing the square root of a negative number—is squared and then added to a second complex number. The sum is squared, and then added to the second complex number, and the process is repeated ad infinitum.

"The point," Arthur emphasized, "is that out of very simple rules, you develop a structure. The structure has certain properties, which mathematically are very beautiful, but they turn out also to have a relationship to some of the things we see in the natural world. Now, the physical world is not self-similar all the way down. At some point it's single atoms, and then it stops. But there's a certain range over which it is self-similar. For example, if we look inside a crystal, again we see more crystals. They have a similar structure to the bigger crystal.

"A bush follows a rule for branching. It grows upward, a branch goes off, then branches go off that branch, and further branches go off that branch. It's similar at every scale, from big branches to medium-sized branches to little branches to tiny branches. It's very similar in this way to the mathematics that we just did. Plants in nature have branching rules that show a self-similarity and a kind of ordered irregularity. If you look at trees, you recognize that there's some order there, but it's not perfect. How do we describe that imperfect order that is both chaotic and organized at the same time? This new mathematics is designed to operate directly in that borderland of imperfect perfection." Arthur illustrated this with a computer graphics animation that used basic rules for branching and development to emulate the growth of a single plant, then multiplied it into a lush garden.

"Let me summarize some of the things we've learned. We've learned that forms can arise without an outside cause, through the application of a simple rule and competing processes like reaction and diffusion. We call this self-organization. We've also seen that there are different kinds of forms. Some of them are unchanging and static, but some of the most interesting ones are growing and changing. Also, we recognized that the most interesting forms are not perfectly ordered. They have irregularities, a little chaos built in. This ordered imperfection often has the mathematical property of self-similarity, and depends very sensitively on the initial conditions. Too much sun or little sun, lots of water or little water, variations in the nutrients in the soil lead to an infinite variety in the expression of these same rules in plant growth. The same rules with slight variations produce many, many different types of forms. For many of these systems, it's very important to recognize that there is a flow of energy or nutrients into the system. We call these open systems. The generation of forms requires that kind of inflow."

In wrapping up his presentation, Arthur laid several questions on the table that linked to larger philosophical issues and would influence our later discussion. "We can now ask, how are the initial conditions set? It could be through contingency, a kind of randomness. We could also ask ourselves, how far are we beyond reductionism? Is this a real top-down approach, or are we still using a quasi-reductionist approach? And I can't help but point out the question of interactions. We could think about the role of consciousness and how it plays into these forms of physical life. Where and how might sentience, or the vital energies that Buddhism refers to as *pranas*, interface with these physical laws and forms?"

Before moving on, we should take a closer look at the notion of emergence. Self-organization goes hand in hand with the arising of emergent properties. As we have said, the term "emergence" describes the onset of novel properties that arise when a higher level of complexity is formed from simpler components. In other words, emergence happens when the whole is more than the sum of the parts.

Emergence has been an active field of inquiry in the philosophy of science for a long time.[5] Both the early British emergentists and modern authors often base their arguments on examples from chemistry. They observe that the properties of water, which is composed of hydrogen and oxygen, are not present in those elements and therefore can be considered novel, emergent properties. When a protein folds, it becomes biologically active, acquiring emergent properties such as catalysis and binding properties that the denatured form does not have. Soap micelles have novel properties—geometrical compartments, permeability, overall fluctuations—that appear in the collective ensemble and are not present in single soap molecules.

This notion of emergence can be found in all fields, even at the level of very simple geometrical figures. When a tetrahedron is formed from six linear segments, properties arise, such as angle and plane, that do not exist at the level of the simple lines. Likewise, a musical phrase can be considered an emergent property arising from the juxtaposition of single notes. Examples from the social sciences are also often given: individuals form a family, families form a community, and so on. All these are novel, emergent properties that arise only when the more complex structure is formed.

Reflections

Returning to the humble level of epistemology, I would like to highlight two points that will bring us to the threshold of the next chapter, devoted to the question "What is life?".

The first point is one made by Arthur Zajonc: that the most remarkable implementation of emergence is life itself. The separate components of a living cell, such as DNA, proteins, sugars, and lipids, or even the cellular organelles, such as vesicles and mitochondria, are per se inanimate substances. From these nonliving components, life emerges once a given space-time organization occurs, without the need to assume any transcendent or mysterious force.

The second point concerns the two important and complementary notions of self-organization and emergence as they relate to the self-organization of a collective ensemble. This is a concept that was dear to the late Francisco Varela. In one of his last interviews before his death, he raises the example of the anthill:

We can now ask ourselves where this insect colony is located. Where is it? If you stick your hand into the anthill, you will only be able to grasp a number of ants, i.e., the incorporation of local rules. Furthermore, you will realize that a central control unit cannot be localized anywhere because it does not have an independent identity but a relational one. The ants exist as such but their mutual relations produce an emergent entity that is quite real and amenable to direct experience. This mode of existence was unknown before: on the one hand, we perceive a compact identity; on the other, we recognize that it has no determinable substance, no localizable core, or any other construction of social insects. When we observe each single living component in the anthill, we are witnessing disordered movements and we would never say that each individual obeys some ordered pattern of behavior. However, the ensemble of the system is quite orderly, there is a social behavior, precise structures are being built, etc. Now, this is a collective pattern that has no localizable center—there is no point in the anthill or beehive that commands this social behavior.[6]

This concept of self-organization and emergent properties as a collective ensemble, without any organized localized center, is nowadays under scrutiny in cognitive science. Several scientists in this field would now agree that the notion of "I" is an emergent property arising from the simultaneous juxtaposition of feelings, memory, thoughts, emotions . . . so the "I" is not localized, but is rather an organized pattern without a center. This brings to mind, of course, the notion of the nongraspable "I" in Zen Buddhism and Buddhism at large, the "I" as an illusion of the clinging senses. In order to make clear the analogy between this last concept and self-organization and emergence in the case of social insects, let me cite Varela again:

> This is one of the key ideas, and a stroke of genius in today's cognitive science. There are the different functions and components that combine and together produce a transient, nonlocalizable, relationally formed self, which nevertheless manifests itself as a perceivable entity. . . . We will never discover a neuron, a soul, or some core essence that constitutes the emergent self of Francisco Varela or some other person.[7]

An Interview with Matthieu Ricard

We succeeded in sequestering Matthieu just before lunch, which was usually served at Chonor House, a small hotel a short, if steep walk from the Dalai Lama's residence, where the meetings were held. It offers a delightful introduction to Tibetan hospitality with rooms decorated in colorful frescoes of Tibetan scenes, gorgeous carpets, and views of the surrounding mountains. On the terrace, buffet tables laden with steaming dishes competed for our attention with equally inviting conversations that embellished on the themes of the week.

The big pots were fuming on the long self-service table and already emitting a wonderful smell of curry and ginger. All around us on the large terrace were more metal tables where people were seated, still engaged in the morning discussion. I saw the venerable Tzogny and his monks, and closer to the house Steve and Eric and Jon Kabat-Zinn. It was a beautiful sunny day, and there was a good atmosphere. The smell of the kitchen helped. Matthieu sat without smiling at our small metal table, and his eyes left without joy the sight of the kitchen pots to fall upon Zara and me, ready with our recording machine and notepads.

LUIGI: The first question is about the relationship between science and spirituality or religiosity. Since you were a molecular biologist who then passed to a religious way of life, you may be able to tell us whether these two aspects of life, science and religiosity, should be considered as two separate things—or whether you see any way to blend them together.

MATTHIEU: First of all, all depends on what you call religion. I do not like the term "religion"; I rather use the term "spirituality." If by "religion" you mean to believe in something that is either illogical or beyond the scope of your personal knowledge, that sets a fundamental difference between science and religion. Now, you can say that science is basically about knowledge, and that spirituality also pursues a kind of knowledge, but applied to a slightly different domain. In particular, if science—by following Galileo—is whatever can be measured or observed or calculated, then it doesn't fully overlap with spirituality; and then, therefore, science and spirituality don't have any fundamental opposition to each other. In this case, what is slightly different is simply the domain of knowledge, the reason we investigate and also the reason we try to acquire this knowledge. Science basically tries to explain phenomena out of phenomena—the way they appear, their relations, the description and analysis of the way they function—and tries to find laws that explain those interactions.

Spirituality is clearly based on the notion of a personal transformation as a human being, so it has to deal fundamentally with the mind, and with expression of consciousness in terms of speech and actions. There is an aspect of clear knowledge related to how the mind works: whether you can speak about the ultimate nature of mind, how thoughts form, how they chain to each other, how they lead to moods, traits, and ways of being.

Also, whether emotions can overpower the mind, and how you could act upon that and perceive that hatred might be harming yourself and others, and things like that. And then verify all that and do experiments on that, with the same rigor and with the same length of time that scientists use when they investigate natural phenomena. So in that case, provided it's based on experience, provided you have enough rigor, provided you spend enough time and you look in the right direction and then try again and again, and you reach some conclusions that others can verify too, in that sense we can say that spirituality is also a science.

Now Matthieu speaks with emphasis, moving his hands and swirling his big colored sleeves around. He seems to have forgotten the steaming pots with their curry.

MATTHIEU: The fundamental difference between science and spirituality lies in the fact that spirituality is clearly pragmatic. It begins with recognition of a particular human condition, which is a mixture of suffering and happiness to different degrees, and the notion that suffering is undesirable. What are the causes of suffering, and are these causes immanent and permanent? If they are permanent, they cannot be fundamentally changed; alternatively, if they are the result of causes and conditions, they could be altered. How? Possibly, by knowing those causes, finding out antidotes, putting them into action, and then eventually removing suffering, which is the natural aspiration of every human being. Even Western philosophers like Aristotle stated, among many other things, that happiness is a goal in itself. We don't look for happiness for something other than happiness itself. So in that sense, this spiritual search is pragmatic. It is not just a tool, it has a clear direction. And so, for me, I think proper spirituality has the same rigor and wish for knowledge that science has.

LUIGI: From what you just said, it appears as if these two domains are separate from each other. One question is whether they can be blended, whether you can perform science within a more general framework of spirituality, as for example Francisco Varela was doing. He was doing science within the larger framework of Buddhism. The question is whether this is generally desirable and possible. Many scientists experience a kind of schizophrenia of the two domains, and the question is whether you can blend them with each other.

MATTHIEU: Again: I think science and spirituality are separate only if you base religion on illogical beliefs, or beliefs in things that you cannot ever verify. But in the way I have tried to describe science as knowledge of the phenomenal world, and spirituality as what I would call a contemplative science—namely, the knowledge of the mind processes—I don't see any difference at all. Spirituality *is* a contemplative science dealing with natural phenomena, a way of studying the mind.

Sometimes the two domains overlap. For someone who is engaged in contemplative science, it doesn't matter what the temperatures of the stars are or how far they are from us. It is true that it is interesting, it's nice information, but it doesn't have any use. In the same way, knowing how emotions arise in the mind doesn't help the astrophysicist to calculate the temperature of a star—they are different domains. They overlap when we are concerned with the same object of knowledge, like reality—what is the nature of reality—a question that is of use for quantum physicists or for others.

LUIGI: It becomes philosophy, in a way.

MATTHIEU: Not only philosophy. If you look at the Copenhagen interpretation, you cannot distinguish the analysis of reality from the quantum physics itself. Or when you are in the field of cognitive science, namely studying the mind, emotions, and pure cognition, then this is completely part of what spiritual science will do. And even more, when you come to branches of neuroscience dealing with emotions, with plasticity of the brain, then you are completely in the same field. Then you can also learn so much from each other.

Now all eyes of the people at the tables look up toward the roofs: there is a monkey jumping from one house to the other, probably made nervous and excited by the good smell of the kitchen. Will the monkey come down and try to steal some food, as occasionally is the case? Matthieu also looks up, but not for long: he is saying something important and looks at us with intensity.

MATTHIEU: So first of all, there is no fundamental difference. The goal is knowledge. Just the domains are different. You cannot study everything at the same time, so either you study natural phenomena or you study the mind, that's up to you. Like physics and geology and knowledge of plants are not contradictory; you simply study different things, but there's no fundamental difference. Then the rigor should be there, the scientific method should be there. And the two domains often overlap—when they have a common preoccupation.

Then Matthieu concluded with another large gesture of his maroon robe.

MATTHIEU: I would say the only thing that spirituality really adds all the time is the pragmatic desire of personal transformation toward a better human being. That's fundamental to spirituality. It's not a failure of science that science doesn't provide ethics, a way of life, a meaning of life, be-

P. L. Luisi interviewing Matthieu Ricard.

cause science was not designed for that. It's not a failure; it's just a natural limitation from the beginning.

ZARA: A lot of the overlap that we've been looking at in Mind and Life, and in the work you've done, for example, with Richie, is looking very specifically from the scientific domain toward contemplative practice as an object of study. Could you envision pragmatic reasons from the Buddhist side to use this information?

MATTHIEU: So far, the interest from our Buddhist side is the fascination of observing correspondences between what we do and what is happening in the brain. This very intense curiosity is not teaching us much about how to improve our method of meditation, of investigating the mind. It might in some areas—there was a hint of that in the case of visualization, because it's the same area as the visual area in the brain. In particular, we were interested in whether there was some kind of conflict between clearly seeing something outside and a very clear visualization inside. Even this is not going to change the way we meditate in our hermitage, at least so far. However, there can be a contribution of science if it turns out that long-term mind training can really affect the plasticity of the brain, as Richie's preliminary results seem to show. In fact, people in this case would find out that meditation is not just a nice time to relax. It is rather that, over the long term, slow changes, one after the other, affect your mind and change your temperament. This is also interesting in the debate on the genetic influence on temperament and acquired transformation [changes gained, e.g., through meditation].

LUIGI: A different question that also has to do with our meeting: in my own dialogue with religious people, I've found that it is possible to find substantial agreement on the question "What is life?"—even Catholic priests accept the idea that cellular life may have originated from inanimate matter. Where I did find difficulties was about the concept of death. For scientists, death means loss of memory of the self: the living structure becomes molecules again, which are reshuffled, and there is nothing left from the previous living being. My Christian friends object to this molecular scenario with the notion of soul. . . . I wonder whether from the Buddhist side this is also so, that you would refuse the view that with death all comes back to molecules and there is no memory of the living creature.

MATTHIEU: The problem of death, of continuity, of reincarnation or life after life . . . they all come down to what is the nature of consciousness. It's much less exotic to speak of that than speaking of reincarnation. If consciousness is nothing else but the brain, there is no need to think about that. Then you come back to Epicurus, who says that it doesn't matter because until we die we are alive, and then when we die we are no longer there . . . there is no problem. The real issue with the above-mentioned items is whether we are going to speak about a beginning, namely [to say] that there is a beginning to things in general, to the living and/or to the universe, or whether we accept the notion of beginninglessness of everything—not only matter but everything that makes the universe—that means also life. Is it the more sensible approach to try to imagine a beginning to everything? Later on I'm going to speak about this point, that basically, nothing cannot become something. In that case, if you take the Buddhist view of beginninglessness, with no prominence to life or matter, then you do not say that life will have its beginning from matter. There's no later stage, there is just beginninglessness. Therefore, both have to have been there in their full array from the beginning, in infinite billion-fold universes.

This was a hard point for me, as a simple-minded scientist, to take in. I made a sign that I needed a few moments of thought digestion. Matthieu's eyes slipped back to the fuming pots. Some people were already helping themselves, and the good smell was breathtaking.

Then Matthieu went bravely back to his line of thought.

MATTHIEU: Therefore, if that is the case, then you go on to consider the Buddhist notion of consciousness and in particular the interdependence between consciousnesses and the inanimate, or the animate and the inanimate, whereby this relation is not a dualistic Cartesian opposition but just part of the same interdependence. By this I mean that they have all the same nature; they are not two distinct things but have different qualities—some have the faculty of knowing, some do not, like stones. That is where Buddhism and science may be a little bit not in agreement for the time being. The Buddhist notion is that things can associate in a continuum with different supports or with different sophistication, like cells and then higher organisms and then eventually intelligent human beings. But that is not automatically and exclusively bound to be with this support. Therefore, there can be a continuum on its own, which at times is associated again and again with different supports, but not necessarily. In that case you can imagine all kinds of other scenarios for what we call reincarnation.

LUIGI: Okay, but this is practically tantamount to a negation of the science view that all comes back to molecules and there is no memory of the human.

MATTHIEU: That's a negation of the science view that consciousness is nothing but molecules. Yes. The association of consciousness with the brain is a phenomenon within those continuums of consciousness.

LUIGI: But when the brain disappears because of death, then the consciousness disappears.

MATTHIEU: Buddhists would say no, because the consciousness is not exclusively linked to that.

LUIGI: That's perhaps the most important point of difference.

MATTHIEU: And the point of difference is also based on the notion of origin. It's linked very much with that. If you have a beginning to things, then you have also an order. Something was there before something else. Then the scenario is the small history we see of our big bang, and of course consciousness came later. But if you see beginninglessness, there's no later, no before. So in that case, it's much easier to think that this notion of a continuum is not totally unreasonable.

LUIGI: Usually people say that in Buddhism all is based on consciousness, so consciousness comes first and then all the rest.

MATTHIEU: No, I don't think that's correct. This reduces it to Cartesian dualism: to say everything is matter is to say everything is solidly, truly existing. Our way to reduce the Cartesian dualism is to say, well, they share the same unreality, they are equally unreal, but with a different quality. It is like in a dream, where you can have a stone that is unthinking, and you can have something that has a quality of thinking even if it's unreal. But there is a different quality. Lack of inherent existence doesn't prevent different qualities.

LUIGI: This no-beginning is also what characterizes, for example, the Indian way of thinking. And on the contrary, in our Christian world, all is based on this notion of creation, of beginning. . . . Where do you think this difference comes from, that we have in our Christian civilization this fixation with the origin?

MATTHIEU: Christian cosmology has not only this notion of creator but also that God created matter first and then he put life to it. So then it's in people's minds. But I think the logic of it isn't very good.

ZARA: And you talk about a temporary association between consciousness and matter?

MATTHIEU: It's an on-and-off interdependence. The interdependence is never broken, but the association is not specific beyond the duration of this life. When the consciousness is not associated with the brain, it doesn't mean that it's disconnected from reality. It's just interdependent in a different way. It's like an on-and-off contact, but nothing can ever be disconnected from anything. Otherwise it would be outside our universe, the relation would stop, and it wouldn't exist for us. Something that is truly independent doesn't exist, as there's no way you can relate to it. So you've got to have an action associated to every phenomenon. Of course it is a question of magnitude, the question of how you detect that action . . . but the coexistence of phenomena implies that they act on each other. Of course there is long-distance action, short-distance action, immediate action, but it's all involved.

LUIGI: Let me also ask the classic question that you've answered millions of times. What was in your personal life that peak experience that convinced you to go the way of spirituality more than the way of science?

MATTHIEU: I was interested in science and then I found that the science of mind, the science of becoming a better human being, the science of giving meaning to life, for me was more interesting than just mapping the chromosome of *Escherichia coli*. Just a personal preference. Then I was inspired by meeting with great teachers, because I saw there was not only an aspiration to spirituality, but there was some living example of what eventually I could become, even if it would take me many lifetimes. Then suddenly it becomes something real. I meet the Dalai Lama, I say, *Okay, that's there*: it becomes suddenly something else than reading it in a book. You see the results of science, and you ask yourself whether you want to go on and add something to that; or you can say, *Okay, here's someone in front of me. What are his qualities and how did he acquire that? Do I want to spend my time trying to achieve what this person has achieved, or do I want to unravel more mysteries of the natural world?* So it's a matter of personal preference. I found that, time being limited, the main point is to do what really matters most for you. What's the point of doing something else? Then you get frustrated immediately.

LUIGI: It sounds almost like you made a rational choice.

MATTHIEU: I made an instinctive choice of what I felt would make it so that every moment of my life is enjoyable.

Here Matthieu burst into laughter.

MATTHIEU: Enjoyable for me doesn't mean being on the beach all day long. It means doing what I'm doing.

Matthieu now looked happy and clapped his hands.

MATTHIEU: The idea of giving up a career and doing something enjoyable doesn't mean just getting a sports car and having the *dolce vita*. For me it meant being with my spiritual teacher, and living the life I'm trying to live. That for me is extremely enjoyable, and I'm not interested in sports cars.

LUIGI: So no regrets?

MATTHIEU: No regrets. Actually, I have a sort of relief that I did that early enough. Imagine if I waited to retire, to be sixty-five years old to start that. What a waste of time! I couldn't start much earlier. You don't leave school to go to the Himalayas, usually. I went there first when I was twenty-one; I left for good when I was twenty-six. I can't complain.

LUIGI: And when you were very young, twelve or so, you didn't hear the voices that people talk about urging you to a religious life?

MATTHIEU: No voices. Rather, reading books and thinking, *This is neat!*

And Matthieu laughed again, looking now more intensively to the fuming pots. He knew that it was time to stop the conversation and go to eat—part of his enjoyable life. "No voices, no voices," he repeated, getting up. "And I did not see any lights. I never saw anything like that. I'm not gifted for that."

Matthieu grinned, and there he went.

3 / Toward the Complexity of Life

IF LIFE ITSELF CAN be considered an emergent property, is it no more then than a particular property of matter? It is now time to consider this question, and it was my task to present to His Holiness what science has learned about the mysterious transition between the world of matter and the world of the living.

I was sitting in the hot seat and His Holiness was looking at me, waiting for my first words. I had rehearsed them many times, but now that I had to

P. L. Luisi trying to explain the origin of life.

utter them they felt inadequate, inappropriate. As many talks as I have given over the years, I should not have felt insecure—but alas, the reality was otherwise. This was not just one more talk on the origin of life, it was a dialogue with the Dalai Lama. I would have to address him directly and look him in the eyes for the whole course of the morning, a good two or three hours. His Holiness was looking patiently at me, his hands folded, intent and sympathetic as always. I found my smile and offered a few words of appreciation, noting how far we had come since my participation in the very first Mind and Life conference, when our interpreter, Thupten Jinpa, was just a young monk. Jinpa, now a very mature scholar of international standing and a family man too, looked a little embarrassed at the laugh this raised from the group, and not least from His Holiness. Then I began in earnest.

"I thought I would begin with some conceptual and philosophical background to illustrate the scientific view about the origin of life on Earth. The terms of the problem are clear. We know that the solar system is about 4.5 billion years old, and that at that time Earth was a fireball, with no life. Then, at a certain point, life arose, most likely through a long process that began about 3.9 billion years ago, when the planet had cooled down. The first cellular fossils are around 3.5 billion years old. From no life to life, in a time window of around 400 million years. Some people say that the time window was much shorter, perhaps only 10 million years. So, how did all this happen?

"The idea shared by the majority of scientists is that life on Earth originated from inanimate matter through a series of spontaneous reactions, which brought about an increase of molecular complexity. This theory was first proposed by a Russian chemist named Alexander Ivanovich Oparin, who produced a very influential book titled *The Origin of Life* in 1924. The date is important. We are in the aftermath of the Russian Revolution, at the climax of Marxist materialism. In fact, Oparin's book was edited in the publishing house of the Workers of Moscow, and has, as the very first line of the first page, the famous communist invocation 'Proletarians of the world, unite.'

"This gives you an idea of the cultural and philosophical atmosphere in which the book arose. However, the view discussed in this book, that life originates from nonlife, is the basis of modern science. And this goes together with the continuity principle, according to which there is no qualitative discontinuity between the inorganic world of rocks and minerals and the organic world of plants and animals.

"The opposite view, voiced by the famous astronomer Sir Fred Hoyle, is that life is due to an accident, a highly improbable event that is comparable, he said in a very famous metaphor, to the fortuitous assemblage of an airplane caused by a hurricane whirling through a junkyard."

"The big bang is a kind of accident, isn't it?" the Dalai Lama interrupted. "If that is the case, what's wrong with life being an accident?"

"We see the reason when we calculate the probabilities. If you calculate that the probability of a given event is 10^{-100}, which approximates zero, then you can say that this event does not have a chance of happening. This is roughly the same probability that a piece of iron would become a piece of gold. In science we would say that such an event does not take place.

"Modern authors in the field claim that the science of the origin of life must adopt the continuity principle, or else one cannot proceed scientifically. This may sound trivial, almost naïve, but in a way, it is not. If you stick to the traditional definition of science, then science is that portion of human enterprise that attempts to explain the phenomenology of the world in terms of the laws of chemistry and physics. If you want to do science, you have to work within the laws of physics and chemistry." (This means, of course, that miracles are out. You cannot be a creationist and a scientist at the same time.)

Determinism and Contingency

Once we accept that to understand the origin of life we have to follow the laws of physics and chemistry, we still face the question whether the occurrence of life was an obligatory process, governed by a causal determinism, or whether instead it was due to the casual vagaries of accidental physical events. The two streams of thought behind these opposite views are the view of determinism and the view of contingency.

The Nobel Prize winner Christian de Duve states that life was a compelled process on Earth: "It is self-evident that the universe was pregnant with life and the biosphere with man. Otherwise, we would not be here. Or else, our presence can be explained only by a miracle."[1] His is not an isolated voice, but is echoed by other important scholars. This view has been strongly criticized by those who argue that this kind of sentence is not science, but rather a statement of faith—and faith is fine per se, but should not be confused with science. However, there are other scientific and cultural movements that share a deterministic view. For example, the adherents of the anthropic principle also state that life on Earth was an inescapable event, and many scientists involved in the search for extraterrestrial intelligence carry the implicit faith that life in an advanced state of development exists on other planets. All these movements, characterized by the belief that life is an obligatory process, share the assumption that life is a sacred property, ascribed to a transcendental design and therefore assuming a designer. Perhaps we should adopt the term "crypto-creationists" for them.

The opposite view is based on the notion of contingency, and it was this I wanted to explain to His Holiness. I continued, "The concept of contingency differs from that of chance or randomness. Contingency is when a large number of independent factors, which are not correlated to one another, simultaneously take place and affect the outcome of a certain event. Each of them can actually be deterministic, but their interplay is so complex and cross-linked that the result of this wild combination becomes unforeseeable.

"For example, a car accident may appear as a chance event, but on a closer look it depends on how fast the driver was driving, on the state of the tires and of the road, whether the driver had drunk alcohol or was in a particular mood, and so on. All these many, many factors conjure up the accident, and the same can be said for a crash in the stock market, a weather catastrophe, and so on. This is contingency. And this view of contingency is central to our ideas on the origin of life and on biological evolution.

"The late Jay Stephen Gould, one of the clearest advocates of contingency, on considering the origin of life and the first steps of evolution, wrote: 'Run the tape again, and the first step from prokaryotic to eukaryotic cell might take twelve billion instead of two billion years.'"[2] He is say-

People of Mind and Life: Roshi Joan Halifax, the Zen master who has her *dojo* in Santa Fe, was one of our number.

ing that the event that led to the origin of multicellular organisms might have happened much later—or never—depending upon the contingent conditions. The implication is that even humankind itself might not have arisen at all. This echoes the familiar statement from Jacques Monod, that we are probably alone in the universe. Monod also said, 'We would like to think ourselves necessary, inevitable, ordained for all eternity. All religions, all philosophies, and even part of science testify to this unwearyingly heroic effort of mankind, desperately denying its own contingency.'"[3]

Was it clear, I wondered, that contingency has nothing to do with miracles? It involves a series of independent steps and events, each one regulated by the laws of chemistry and physics. The final event is accidental but not miraculous: it is just the complexity of the interrelations among all these steps and events that make the overall result unpredictable or even stochastic in nature: it may happen, or not.

Panspermia

Up to this point, we had looked at some of the conceptual background underlying the research on the origin of life. The great majority of scientists share the view that life on Earth originated from inanimate matter, from the simplest molecules produced under prebiotic conditions. I continued:

"In addition to these molecules that were formed on Earth, we had a lot of molecules coming from the cosmos in prebiotic times, and this is still happening now. Every year several tons of cosmic dust come from space, and this dust contains a lot of organic molecules. Meteorites and comets bring additional compounds. There is a theory according to which the water we have on Earth mostly came from meteorites impacting against the Earth. All this led some scientists to ask: 'If so much material comes from the cosmos, why not life itself?' This is the notion of panspermia, which is still occasionally voiced in the scientific community. The idea is that the seeds of life came from elsewhere."

The Dalai Lama interjected, "That is not really a solution. You just push the question further."

"Yes," I agreed. "One shifts the question of the origin of life to another unknown planet under unknown conditions, and this does not help to solve the problem. The majority of scientists agree on that. However, behind panspermia lies the idea that life is an all-pervading property, that it's spread all over the universe, and then it pops up on different planets."

His Holiness persisted, "So from the perspective of these theories, can it be also maintained that there was life before the big bang?"

"No. We don't take panspermia seriously, although some people still consider it an important view." It actually makes no sense to pose the question of life before the big bang. It does not even make sense to talk about the laws of physics or even about time or space before such an event.

It seemed wise to tackle a more fundamental question.

What Is Life?

"I thought it would be proper now to talk about how modern science defines life. What is life?

"There are different scientific schools of thought about what life is. One focuses on the cellular level, another on molecular replication. I will stick to the cellular view for now, because this also permits me to explain the work of Francisco Varela, who, with his theory of autopoiesis, outlined together with his teacher, Humberto Maturana, can be regarded as the main proponent of this school of thought. Since 'life' is a very broad and vague term, the scientific approach to its definition is to look for where life has its simplest expression. This place, for science, is at the level of microbes and unicellular organisms. So we'll start with these tiny and simplest of living things.

"If a biochemist zooms into one such simple living cell, he discovers enormous complexity. Don't be scared, but that diagram represents the enormous biochemical complexity of a simple cell. Each point represents a compound, and each line represents a chemical reaction. Each of these many thousands of chemical reactions is made possible, or catalyzed, by enzymes, which are very large proteins. There are as many enzymes in one cell as there are reactions.

"Each tiny cell contains thousands of compounds, which are all interrelated in an extremely large complexity. In life on Earth, there is no single-celled organism that has less than a few hundred genes, which means thousands of components, and all these components are related and talking and linked to one another in an extremely complex maze. If even a tiny cell is so complex, what can a scientist do?"

The complex maze projected on the screen seemed to trigger a long discussion in Tibetan between the Dalai Lama and the Buddhist scholars at his side. Out of this surfaced a question from the Dalai Lama: "Considering the question of the transition from life to nonlife, when you have a tree that dies, for example, and it goes back to minerals, what becomes of the life? Likewise, in terms of minerals becoming life, how does this transition take place?" His Holiness had jumped ahead to the very question

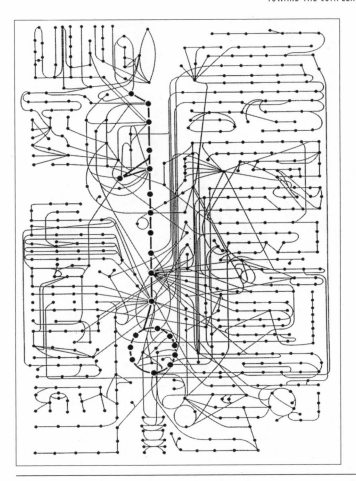

The complexity of the metabolism of a cell. Each point represents a chemical compound, each line a reaction; and each reaction needs a large catalytic protein (an enzyme) to occur.

that I had anticipated would be central to the dialogue to come. I was gratified, but I wasn't quite ready to go there yet. There was still much to cover to lay the foundation for this discussion.

"We'll come back to this question later on, but to answer in a nutshell, science holds the view that when something dies, it dissolves into its molecules, which are reshuffled. The molecules are taken and utilized again to make new living—or not living—things. So science also has this concept of a wheel of life. Except that for scientists, when something dies, it really dies, in the sense that all memory of the particular self of the individual is lost."

His Holiness asking a question—always a challenge for the scientist.

The side discussion in Tibetan was continuing. Thupten Jinpa alerted us to what was going on: "There is a problem of linguistics here, in part. The Tibetan term that is used for 'biological organism' is a newly constructed term that literally means a phenomenon that has some form of life, some process of growth or coming into being." It was obvious that the conundrum held some humor for His Holiness, as the side discussion with his translators continued with much laughter.

His Holiness addressed the meeting again with his question formulated as follows: "The issue can be seen just in the example of a blade of grass: it's born, it's generated, it goes through a process of radical transformation even in one year. Then it's no longer grass: it's transformed into something else. A tree also, perhaps over a century, transforms into something else. But if we look at a longer time span, like a million years, then even a stone or a hill can undergo radical transformation. So again, where exactly is the demarcation from the inorganic to the organic? It can't be simply the process of change, so it has to be something else. What exactly is that 'something else' that differentiates what is living from what is not living?"

Eric Lander stepped in with a further effort at clarification: "I think the question is, why is a stone, which undergoes transformation, not alive? Where do you draw that line? Why do you consider the tree that undergoes transformation alive, but the mountain that undergoes some change not?"

The question only made it clear why we needed an operational definition of life, one that permits us to discriminate between life and nonlife, and so I asked for patience as I developed my theme.

The Essential Features of a Biological Cell

"So we were saying that this biochemical complexity of a tiny cell is enormous. What can scientists do when they are facing this kind of complexity? They build a model. They say, yes, this is very complex, but what is essential? Can we make a scheme that represents the essence of the thing?

"The essentials of a cell can be summarized in a diagram like this, where you have a membrane, a wall that demarcates the inside from the outside, and therefore defines a closed entity. Through this membrane some nutrients come in, which then undergo the series of reactions that we have seen before. There are many, many thousands of reactions in each of your cells at this moment, but essentially, all that is being destroyed by biochemical mechanisms (sugar that is oxidized, an amino acid that is transformed) is rebuilt again inside the cell by other reactions. Or, in Francisco Varela's terminology, 'The cell's main activity is to maintain its own

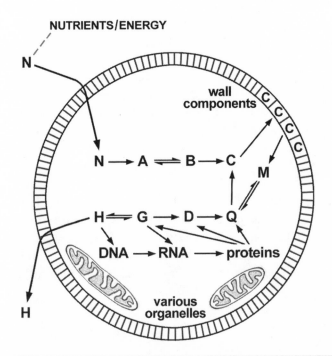

The cell's functioning schematized in a simple diagram, with nutrients N entering through the semipermeable membrane and products H being expelled. There are very many transformation and regeneration reactions inside, but the living cell maintains its own identity thanks to the regeneration mechanism from within.

identity (self-maintenance) in the face of the enormous number of transformations—and the cell does so thanks to a process of regeneration from
within.' From this, we can arrive at an equivalent description that corresponds to an operational definition of life: a system that is spatially defined
by a boundary of its own making and that sustains itself by regenerating
components that are being used up."

I paused to gauge the reaction to these statements, and the Dalai Lama
jumped in with a series of questions: "But what maintains the identity of
the cell as a whole? We have all these internal processes that are arising
and destructing. Is there a continuity of the individual components within?
That is, do you have a lot of individual continuities within that allow for the
continuity of the whole, or does the individual continuity of these separate
internal processes cease? Is there a true destruction happening? And the
identity of the cell as whole—does that change or not change?"

In fact, what I had not stressed enough in my presentation is the link
with the notion of self-organization. This is what permits the maintenance
of identity. A living cell is one of the most sophisticated examples of emergence and self-organization, where in fact the combination of these two
principles acquires a particular quality: the capability of maintaining its
own self through a dynamic network of processes, which can be defined as
biological autonomy. As Francisco Varela would say,[4] cellular life emerges
at a level of complexity characterized by the fact that the product of this organization is the living system itself, and that there is no separation between "producer" and "product." At this point, Francisco Varela would also
have introduced the notion of operational closure, where all the information needed for the operation of the system is contained within the system
and there is no need of external instruction.[5]

I tried to clarify: "Inside the cell, components are destroyed, but they
are remade by these networks of reactions that are localized inside. So it's
a continuous process of remaking the products of its own production. Until now, that's how we have defined life: a system that is spatially defined
by a boundary of its own making, and that is self-sustaining through its
ability to remake its components from within. So the cell is a factory that
makes itself."

Autopoiesis

From simple, basic observations about the life of a cell, Maturana and Varela (often referred to as the Santiago school) arrived at their idea of autopoiesis. This is basically a generalization of the notion of cellular life.

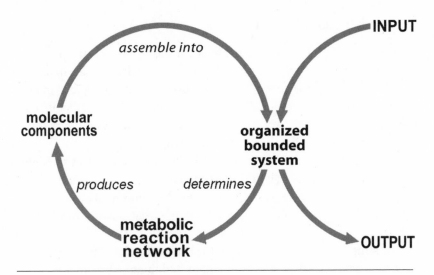

Autopoiesis and the cyclic logic of cellular life. In the cell, which is an autopoietic unit, the boundary determines a network of reactions that in turn produces the molecular components that assemble into the organized system that in turn determines the reaction network . . . and so on, without beginning or end.

The word "autopoiesis" comes from the Greek and means self-production. An autopoietic unit is a system that is capable of sustaining itself through a network of reactions that regenerate all components that are used up in other reactions. This definition does not imply any particular structure. It is valid for a biological cell based on proteins and DNA, but it could also apply to a synthetic system of artificial life. In this way, autopoiesis is capable of capturing the mechanism that generates the identity of the living. The components organize themselves in a bounded system that produces the components that in turn produce the system, etc. In this way, the blueprint of life obeys a circular logic without an identifiable beginning or end. Autopoiesis is not concerned primarily with the origin of life. It is a representation of life as it is, here and now. Although the system is open from the physical point of view, from an epistemological perspective it has a logical operational closure.[6] This characterizes the system as an autonomous identity that can be defined as autoreferential: it produces its own rules of existence. This notion of autopoiesis has been adopted not only by biologists, like Lyn Margulis,[7] but also by social scientists[8] describing a system that maintains itself from within, by the laws of the system itself.

The Green Man and the Farmer

I continued, "The observations made about the working principles of a cell lend themselves to a surprising discovery: that what we say about the tiny cell is also true for an elephant and any other macroscopic living organism. In fact, the operational description of cellular life as a factory that makes itself from within is generally valid for all life, and is the very thing that discriminates the living from the nonliving.

"With my students I use a game that I call the game of two lists.[9] Suppose that an alien, let us call him a Green Man, comes to us to learn what life is on our Earth. In my story, the Green Man finds a farmer on Earth and says, 'Look, I have this list. Can you tell me what is living and what is not?'

"The simple farmer quickly divides the list into two lists and says, 'Look, this is very easy. The fly, the tree, the mule, and the mushroom are living; whereas the automobile, the computer, the crystal, and the moon are not living.' Our Green Man is very surprised by the speed with which the farmer makes this division, and he wants to know how he did it. What discriminates the living from the nonliving? What is the quality that is present in all these living things, from the fly to the elephant? Despite their diversity, they must have something in common. Furthermore, this 'something' should not be present in any of the items on the list of the nonliving: the crystal, the moon, the computer. What is this quality that is the common denominator of the living?

"With my students, it takes a while to arrive at a solution—but not so long. The common denominator of the single living thing is not self-reproduction, because some individuals can be sterile, and for reproduction in higher organisms you need two individuals. What is common to any living thing, be it a tree, a mushroom, or a fish, is this capability of self-maintenance, of remaining oneself despite the many transformations that take place inside. This is also true, of course, for each of us. Each of us loses millions of skin cells every second, but these skin cells are remade from the inside. Most of your hemoglobin is destroyed in a few days, but is remade from the inside. You cut your hair, and this hair is remade from the inside. This feature of regenerating the components, maintaining your self despite this large number of transformations, is the very essence of life as we define it."

The audience seemed satisfied by the story of the Green Man and the farmer, so I continued: "The living cell works as a distributed ensemble, but it does not have a center that directs the action. The many processes and reactions are coupled together by self-organization, but there is no central director. In cognitive science in general, as well as in all modern theories of complexity, this notion of self without a center comes up again and again. This is well known to people who study brain function, like Le Doux, Antonio Damasio, Gerhard Roth, and also of course Francisco Varela. They might tell you, for example, that the mind is an ensemble of activities, feelings, emotions, memories, and thoughts. There is thinking without a thinker. There is a beautiful house, but nobody is home. It's the collective ensemble that accomplishes this self-organization, creating a whole without a center. The same can be said for the living cell: there is life in the ensemble, but no localized center of life.

"Connected with this definition of life is a definition of death. For us scientists, when a fish dies, it gives back all the molecules to its environment. Molecules are again utilized by nature to make other things. All our molecules have been used many, many times before. We may now have in our bodies some molecules that belonged to dinosaurs. It's a reshuffling of nature. In other words, in science we hold the idea that with death there is no memory of the self, and no idea of reincarnation."

I had touched again on a point of contrast between science and religion, and thought it worthwhile to emphasize it, expecting a reaction from the Buddhist camp more intense than before. I added, "I have had many dialogues with priests in Switzerland and we always found ourselves in agreement about the definition and even the origin of life. But as we came to the question of death, the difference between science and religion became apparent." But there was no comment. Good. So I decided to go one step further. To consider life in terms of single cells may be instructive, but not realistic. No cell lives alone; there is always a community of cells, and those communities live in a particular niche, or, more generally, environment, on which their life is dependent.

The Bridge from Cell to Cognitive Science

"Along with the question 'What is life?' there was another question on Maturana and Varela's agenda, namely 'What is cognition?' In investigating the relation between these two questions, Maturana and Varela arrived at the conclusion that the two notions, life and cognition, are indissolubly linked in the sense that one cannot exist without the other.[10]

"The starting point is the interaction between the autopoietic unit and the environment. The living unit is characterized by biological autonomy and at the same time is strictly dependent on the external medium for its survival. There appears to be a contradiction here, and life must indeed operate within this apparent contradiction. The interaction with this environment is always a very specific one, in the sense that the interaction a butterfly has with the environment is different from the interaction that a worm or a human being has with the environment. Maturana and Varela call this particular type of interaction 'cognition.'

"Varela[11] recognized that the choice of the term 'cognition' was not ideal, as it has a strong anthropomorphic connotation. One thinks immediately of human cognition. According to Varela and Maturana, however, there are various levels of cognition, including those at lower degrees of life's complexity: from unicellular to multicellular organisms, from plants to insects and fish and mammals, each with its own type and degree of cognition. Each corresponds to a different level of life's complexity. Still, cognition is a notion that applies only to living entities, and not to the inanimate world—chemical recognition is not cognition."

The environment is independent of the organism; it does not prescribe or determine changes to the organism. It induces a reaction, but the changes are determined by the internal structure of the organism itself. It is the structure of the living system and its previous history of perturbations that determine what reactions the new perturbation will induce. Changes, mutations, and evolution are seen as the result of the maintenance of the internal structure of the autopoietic organism. Since the dynamic of the environment may be erratic, the result in terms of evolution is a natural drift, determined primarily by the inner coherence and autonomy of the living organism.

Cognition and Enaction

Cognitive interaction is the result of the internal structure of the autopoietic system. In this sense, the view of Maturana and Varela is the opposite of the representational model, according to which cognition is primarily the act of taking a picture of the external environment. In the autopoietic view, cognition is instead a mutual interaction between the inner structure of the system and the environment, in the particular sense that the environment is "created" by the sensorium of the living organism during the interaction itself.

The term "create" may sound far-fetched, but often it has a clear physical meaning: for example, the onset of photosynthetic organisms may have created a novel, oxygen-rich environment. The spider may create a web or a beaver its woody construction, not to mention the cities and freeways of people. More generally, however, this term denotes the particular and, in a way, subjective coupling that makes living organisms and their environment compatible and in harmony with each other: the living structure and the environment "create" themselves reciprocally.

Varela[12] coined the word "enaction" to indicate this very process of mutual calling into existence: the organism with its sensorium "creates" its own world; the environment allows the living organism to come into being. Thus the term "enaction" indicates a mutual process of adaptation as a coemergent act. Actually, the term "coemergence" is an appropriate alternative to "enaction." The process of coemergence is simultaneous and equivalent to that of cognition, and all of this is equivalent to the process of life. Again the emphasis and the overall concern of the enactive approach is not to define cognition in terms of an objective relation between a perceiver and a world, but rather to explain cognition and perception in terms of the internal structure of the organism.

In this way, as explained above, cognition is defined at various hierarchic levels: in micro-organisms, then in multicellular organisms, and so on,

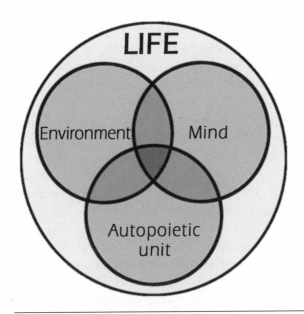

The embodied mind, in a simple pictorial representation showing the trilogy—but actually the unity—of body (the organic autopoietic structure), environment, and mind perception (cognition).

each time acquiring a higher level of complexity and sophistication. At the level of mankind, cognition may become perception and consciousness. But the same basic mechanism is operative at all of these levels. In the process of enaction, the organic living structure and the mechanism of cognition are two faces of the same phenomenon, "the phenomenon of life."

It is at this point that the theory of autopoiesis creates a bridge between biology and cognitive science, and in this way the very notion of consciousness is seen as deriving from the autopoietic organization of life at the level of the human being. Accordingly, one does not exist without the other: there is no organic human life without consciousness, and there is no consciousness that is not embodied in the organic life. Human life is the coemergence of the two.[13]

* * *

At about 10:30 each morning, the Mind and Life conference takes a tea break. The announcement of the break is greeted with a buzz of voices and movement. The participants move out into the garden, where coffee, tea, and cakes are waiting, displayed on a long table covered with a white cloth. The restroom at the end of the long corridor still has the same unreliable plumbing as in previous years, and we have come to accept this fatalistically as a fact of life. Outside, under the bright sun of India's early October, an atmosphere of friendship and lightness surrounds the table. The cakes are much improved since the steely biscuits of the earliest conferences, the tea is tasty, and the coffee is of course Nescafé®, a great luxury over here.

The older monks are also with us, and the discussion on the themes of the early morning continues, distributed among various small and heterogeneous groups. His Holiness does not join us outside; he usually stays in his seat through the break. A few people press close around him, and he gives them his attention generously. This is the time for the scientists to ask more personal questions, and the opportunity also for the observers—those who sit in the second and third rows—to make a personal connection. Some are returning scientists who presented at previous Mind and Life conferences, others are learned Buddhist scholars from India or Europe or sponsors who have supported the Mind and Life Institute. These discussions around the Dalai Lama's chair during the breaks are extremely interesting, and always present a dilemma for the participants: whether to stay and listen or head for the sun and cakes outside, where the discussions are also exciting. It is indeed a time when you would like to be in at least three different places at once.

But several people were asking questions about my presentation, and Matthieu Ricard also approached. I complimented him on his recent photography book[14] on Buddhism in the Himalayas, the most beautiful book I have seen on the subject, both for the quality of the pictures and for the intelligence of the text. He was pleased, in his modest way, and then turned to the subject of my talk. He said he had arguments against the story that science presents on the origin of life, and would offer a Buddhist view. I voiced my assumption that the Buddhist view was probably the reverse of the science view: instead of starting from inanimate matter and progressing through molecular evolution to the cell, to humankind, and then to consciousness, wouldn't the Buddhist begin with consciousness and then work down to matter?

Matthieu shook his head. "No. One does not have to premise a beginning of any sort," he offered, "be it matter or consciousness."

I had doubts: "Don't you have to start somewhere?"

He smiled mysteriously and told me I would have to wait for him to talk about the notion of beginninglessness with the group as a whole. This left me very curious and uncertain. I wanted to ask more, but Matthieu had been already captured by another small group. I had to wait. . . .

The bell rang to resume the session, and we all returned to our seats with pleasure and excitement.

Experimental Approaches

Until now we had approached the question of the origin of life from a philosophical point of view. We had also settled on a definition of life, based on the observation of cells. Now it was time to consider the experimental aspects of the question. Can Oparin's proposition, that life on Earth originated from inanimate matter, be tested experimentally? In particular, can one reproduce in the laboratory those stages of increasing molecular complexity, from molecules to macromolecules and from there to the first metabolism, to arrive finally at structures capable of self-reproduction? Let us start by looking at the operational principles utilized by the experimentalists working in the field of the origin of life. What are the ideas guiding them?

The first postulate, that life originated from chemical reactions on Earth, leads to the second (and third): that it should in principle be possi-

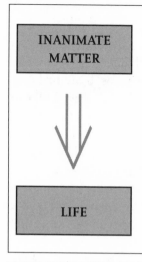

1. Life originated from inanimate matter as a spontaneous and continuous increase of molecular complexity. Chemical continuity principle—no transcendental principle.
2. The chemical process(es) to transition to life can be reproduced in the laboratory with the presently available chemical techniques and chemicals.
3. This can be implemented in a reasonable (hours or, at maximum, days) experimental time span—once you know the right combination of prebiotic compounds and the necessary conditions.
4. Since there is no documentation on how things really happened, there is no obligatory research pathway.

Main assumptions generally accepted by researchers on the origin of life.

ble to reconstruct this pathway in the chemical laboratory and therefore to "make" some primitive form of life. Since we do not know how nature really did it, as there are no fossils from the time before the first cellular fossils, each scientist is free to use his or her own creativity in order to reconstruct—conceptually and/or experimentally—this pathway, provided that the means used are prebiotic.

Aside from the experimental difficulties, is the construction of this pathway really possible? Aren't there conceptual difficulties? I tried to translate this into more concrete terms: "The branch of science that is concerned with those questions is called prebiotic chemistry. Its beginnings go back to a young American chemist, Stanley Miller, who in 1953 published a very important paper in *Science* that at first sight seemed to give a positive answer to the possibility of reconstructing the pathway of the transition to life in the laboratory.

"Stanley Miller decided to put what he considered to be the main components of the prebiotic atmosphere—hydrogen, ammonia, methane, and water—into a flask. And then he let an electric discharge go through this flask to simulate lightning as a source of energy. To his surprise and the surprise of the scientific community, he found that the flask turned brown, due to compounds such as amino acids and other rather large molecules formed from those simple gaseous components.

"Now, this experiment is very important because it shows that the building blocks of life, amino acids and sugars, can be formed from very simple molecular components with a very simple energy source. In addition, it answers the question why these molecules, amino acids, form and not others. They form because they are the most stable of the possible compounds."

Following Miller's experiments, several chemists begun to study what they called prebiotic chemistry, and this branch of science was very successful, in the sense that many compounds were synthesized under simulated prebiotic conditions. Even the aromatic bases of the nucleic acids, like adenine and cytosine, were created in this way, as were sugars.

Self-Replication

Prebiotic chemistry is one direction that the study of the origin of life has taken. However, much more research is needed before we reach the complexity of life mechanisms. Another direction is the study of chemical systems that are capable of self-reproduction without the help of enzymes. Self-replication (or self-reproduction—the two terms are often used synonymously) is a fundamental process in all mechanisms of life, although, as we have emphasized in the discussion of autopoiesis, self-reproduction cannot be considered the cause of life. Chemists try to mimic the self-reproduction process in the laboratory with simpler systems, just to show that there is nothing mysterious in it, that it is just a chemical process, even if not the simplest. I tried to explain:

"Self-replication is usually a nonlinear process from which one element produces two, from two you get four, from four you get eight. This is a very powerful mathematical apparatus that nature developed.

"Let me give you an example. If you have a normal chemical process by which A becomes B, and this step takes one second, then you need two seconds to make two molecules of A, ten seconds to make ten molecules, and so on, linearly. To make, say, 100 grams of a molecule having molecular weight 100, you would need to make the Avogadro number of molecules, 10^{26}, for which you would need 10^{26} seconds, a time much longer than the age of the universe.

"If instead you proceed by a nonlinear process, where one makes 2, and from these you get 4, and from these 8, 16, 32, 64, by a self-replication mechanism, you need only 79 seconds to make the same amount of matter, those 100 grams we were talking about." I looked around to see whether the audience was still with me. Arithmetic is not very popular in

Dharamsala. I decided to offer a much simpler example to explain the nonlinearity.

"You know the story of the emperor and his wife who had a dispute that they decided to settle by playing chess? The lady says, 'If I win, I simply want some rice. I want one grain of rice in the first square, two grains of rice in the second, four in the next, eight in the next, and so on.' The emperor says okay. They play, he loses, and he discovers that with this power of self-replication, he doesn't have enough rice in his kingdom to pay her. This gives you an idea of the power of self-reproduction mechanisms in nature."

At this point I showed some diagrams of simple chemical systems capable of self-reproduction, starting with the von Kiedrowski system of the early 1980s and going further to the self-reproducing micelles and vesicles introduced by the Zürich group in the early '90s. It was hard chemistry, and I had the impression that it was too much for my audience. But the important message here was that you can make self-replicative systems in the chemical laboratory, so there is nothing mysterious or magic about self-reproduction. It is something we can understand and make happen in the laboratory.

However, this was all at the level of small molecules or molecular systems. And there was a very important point I wanted to stress here: "Even if chemists had access to all prebiotic compounds of low molecular weight, such as amino acids and lipids and sugars and bases, they still could not solve the problem of the origin of life. Why?"

The Conundrum of Macromolecular Sequences

The point is that life as we know it is determined in large part by very long molecules, the so-called macromolecules or biopolymers, like nucleic acid and proteins, in particular enzymes. These large molecules consist of ordered sequences of amino acids in the case of proteins or of bases in the case of nucleic acids. The term "ordered sequence" is very important. The 20 amino acids in nature are analogous to letters of the alphabet. Only particular linear sequences of letters make meaningful words, and only particular sequences of words make meaningful sentences. Likewise, only particular sequences of amino acids (or bases in the case of nucleic acids) make biologically meaningful proteins. The analogy goes further: even if you consider a very rich vocabulary, the number of meaningful words is a tiny fraction of all possible combinations of letters. For example, if you calculate how many strings of 10 words you can make with the 24 letters of the alphabet, you reach the astronomic number of 24^{10}, which is around 10^{30}, a one followed by 30 zeroes. The obvious consequence is that only a tiny fraction of the potential structures have been implemented on our Earth.

A very important family of proteins are the enzymes, which are the catalysts of life. They are proteins that allow all reactions in our cells to proceed swiftly and efficiently at room temperature. Each enzyme catalyzes a different reaction and is therefore different from the others, and the biological activity of each enzyme is due to its very specific sequence, a precise linear structural order. This sequential linear order determines a three-dimensional folding that in turn determines the biological activity. There are millions of different enzymes on Earth; these structures exist not because they are thermodynamically more stable than the other theoretically possible chains, but rather because they are the products of a long evolutionary history that shaped their sequence in the course of prebiotic molecular evolution.

The construction of this sequential order in the chains poses a fundamental problem in the origin of life. How was it established under prebiotic conditions characterized by a complete lack of regulative mechanisms? It is unthinkable that mere chance could have created all this natural order, as in nature we have millions and millions of functional enzymes and nucleic acids, not just a few. By which mechanisms have all these linear structures been selected out?

Clearly this question takes us back to the controversy between determinism and contingency outlined earlier. Either the proteins existing on Earth are the product of an obligatory causal order, or conversely, if contingency dominates, then the existing proteins are just a sample selected out by random processes, dictated by the contingent conditions existing during their evolution pathway.

The Dalai Lama took an interesting tangent in response to this: "Given this minute proportion of the proteins actually existing on this Earth with respect to all the possibilities, is there any reason to believe that on other solar systems throughout the rest of universe the same proteins would have emerged?" There is no reason to believe this, as I tried to clarify. There is nothing special about our existing proteins from the thermodynamic point of view. The point is precisely that they most probably exist through mere contingency. Of course, once the first forms of life were established, a series of optimization processes began, and the protein structures became more and more consistent with the cellular life processes.

"The ratio between the theoretical number of possibilities and the number of proteins we really have corresponds to the ratio between the size of the universe and the size of one atom. In other words, we are made out of a very, very tiny fraction of proteins: nature has realized only a tiny fraction of the possible."

Having made the point that the proteins existing on Earth have been mostly shaped by contingency, I reached the conclusion that the reconstruction of these particular proteins in the laboratory is not an easy matter—actually it appears to be an impossible task. There is no way, in fact, to know and therefore to repeat in the lab the steps and the conditions that led to the formation of the actual proteins. Each of them has been formed and shaped by a series of unknown parameters during the course of a long evolution. At the most, what we can do in the laboratory is to show that, using prebiotic processes, the synthesis of proteinlike macromolecular sequences is possible in principle.

Is the construction of living cells in the laboratory therefore an impossible task? No, because there is another way to construct simple forms of life, a modern synthetic approach to the so-called minimal cell.

The Synthetic Biology Approach to Minimal Life

In order to understand this approach, we need to first define what is meant by "minimal life" and the minimal cell. The enormous complexity of modern cells—simple unicellular organisms contain thousand of genes and all corresponding macromolecular components—is most likely the result of millions of years of evolution, and is in large part the result of survival in a competitive environment. Most likely the early cells living in a more permissive environment did not need such complexity. Is cellular life possible with a much simpler structure and many fewer components? This leads to the notion of a "minimal cell" that contains the minimal and sufficient number of components to be defined as alive.[15] For simplicity we can say that a cell will be considered living when it has three main properties of life: self-maintenance, self-reproduction, and the ability to evolve.

How can one "make" this minimal cell in the laboratory? "We start with a shell that resembles the shell of biological membranes. The idea is to put inside it the minimal number of enzymes and nucleic acids to permit the construction of a living cell.

"Now, we first need cell-like compartments, and for this purpose we use liposomes, which are tiny spherical bubbles with a dimension comparable

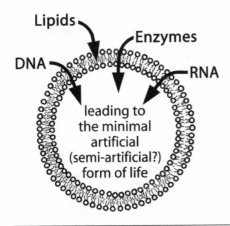

The approach to the construction of the minimal cell, containing the minimal and sufficient number of components to be called alive.

to those of biological cells. It is relatively easy to make such liposomes in the laboratory by putting simple molecules of surfactants, or lipids, in water."

Here I showed some pictures of these liposomal structures, which prompted a question from His Holiness: "Is this a real photograph?"

"Yes. This spherical vesicle is formed by about five billion molecules that assemble together in this perfectly spherical form, a magnificent example of self-organization. Our natural lipids are also surfactants, and in this case, when the vesicle is made out of lipids, it is called a liposome."

"And they naturally form that shape?"

"Yes, just like soap bubbles. It's a kind of spontaneous self-organization. It takes place by itself because this structure is more stable than the single components. Once we have these liposomes, we can move one step further."

We have found conditions under which liposomes can self-reproduce, so that their population increases at the expense of smaller molecules. The relevance of their self-reproduction lies in the fact that they are good models for cells. In other words, self-reproduction of liposomes simulates self-reproduction of cells. Of course, biological cells self-reproduce with all their internal genetic content, whereas liposomes are just shells. But still, it is an important first step."

How is this self-reproduction of liposomes possible, chemically speaking? "I will spare you the chemistry, but basically we add some simple binding chemicals—the so-called precursors—to the liposome membrane.

At this point a very simple chemical reaction called hydrolysis occurs, splitting the precursor molecules into the lipid molecules that form the liposomes (or vesicles). Thus, this original vesicle grows, and by growing it becomes unstable, and, after reaching a certain instability, it divides up into smaller vesicles."

Self-Reproduction of Model Cellular Systems

"The next step is to put biological macromolecules into the self-replicating liposomes." I showed an image from one such experiment. "This is a vesicle or liposome in which we have put one enzyme, called Q-beta replicase, which is capable of making copies of RNA as well as several other components. While the vesicle self-replicates by the chemistry I explained earlier, on the inside there is a replication of RNA through this enzymatic system. So we have a kind of semisynthetic system in which the shell and the core both replicate.

"Now is this life in the lab? Not yet, because as the self-reproduction keeps going from liposome generation to generation, the enzyme and the other initial components dilute out. So after a certain number of replications of the liposome shell, you will find a lot of liposomes without the enzyme, or without RNA, and very few or none will contain all the ingredients to make copies of RNA inside one given liposome.

"However, you see that this is already going in the direction we want, and there are other experiments of this kind. For example, a few proteins have been expressed inside liposomes.

"I have occasionally been addressed by journalists who ask by telephone, 'Have you made life?' The answer is no, but in fact there is the expectation that within a few years, people will be able to make semisynthetic living cells in the lab. This will be a kind of life, but much simpler than that of living cells. These simple living systems can be considered as models for protocells, the first early cells from which our life and cell evolution started.

"I finish by reconnecting to the initial question, 'What is life?' We have seen that this question can today be approached and answered in terms of chemistry. Molecules and chemical structures alone cannot, however, bring us to the understanding of life. We need at least the approach that Maturana and Varela offer, with the notions of self-organization, emergence, autopoiesis, and enaction, as mentioned before. Let me repeat, in concluding this part, that an autopoietic unit is a structure or system that is capable of self-maintenance due to internal regeneration of components.

The term applies to a biological cell, but also to any living macroscopic organism, and can apply to noncoded life as well. We have already argued,
following Maturana and Varela, that there cannot be life without the interaction with the environment, which leads to the concept of cognition.
There is a mutual integration, a mutual calling into creation, that Varela
calls enaction (and that I, and other authors, call coemergence). In this
sense, life is equivalent to cognition; one cannot exist without the other.
This brings us to a unity between body and mind, which is the legacy of
Francisco Varela. In my own work, I'm trying to convey this message,
namely that the biochemical basis of life, although based on molecules
and their interactions, is not only consistent but also one with the cognitive domain and the spiritual domain."

There was finally a moment's pause as I concluded. I looked at His Ho-
lines, who was looking at me with attention, his hands crossed over his
legs. I looked then at the Karmapa, who returned my gaze with a friendly
smile. I still wondered how much he was able to follow of these scientific

Lunch served at the Chonor House. Good food, but mostly good conversation
companions. Here are Matthieu Ricard and his mother, conversing animatedly
with Richard Gere.

arguments. *I must really talk to him and ask; I am very curious about his way of understanding.*

A Philosopher's View on Emergence

The notions of self-organization and emergence have been classic themes in the philosophy of science for many years, and it was not to be expected that our resident philosopher would let all this glide past without adding his own thoughts. Like many arguments in the philosophy of science, emergence can be considered on two different levels. Scientists favor an "epistemic" interpretation, focusing on knowledge of observable patterns and empirical reality. Philosophers, on the other hand, lean toward an "ontic" interpretation, theorizing about things as they really are, independent of any observational or descriptive context. With this in mind, here is Michel Bitbol's contribution on the fiercely debated concept of emergence, which applies to many natural phenomena, as well as to the properties of life and mind.

Michel began by framing the essential question: "Can we say that life is a radically new property, or state, or process, arising from nothing other than a large number of interacting atoms and molecules, none of which can be said to be alive when taken in isolation?

"As you can see, the problem is formulated in heavily ontological terms. Those who raise it want to know whether such emergent large-scale properties truly exist; they want to know whether these large-scale properties are more than just epiphenomena, whether or not they have the causal power of altering other properties.

"This ontological formulation of the problem of emergence is not surprising in view of its historical motivation. The concept of emergence was invented in order to find a satisfactory compromise between two extreme views. The first of these two is monist and materialist: it says that nothing exists other than material elements and their properties. The second view is dualist: it says there are two substances or two realms of being: mind and matter, or life and inanimate matter. Emergentism aims at finding an ontological 'middle course' between these two views."

Emergence and Reductionism

"Of course, a middle path may still show a slight bent toward one of the two extremes," Michel noted. "Nowadays, the most common bias is materialist and reductionist. The appearance of new and autonomous features is easily explained by ontological reductionists by assuming that our experimental or perceptive analysis is coarse-grained. After all, if our senses or instruments are a little bit coarse, it's no wonder that large-scale behavior looks new and autonomous compared to elements studied at a much higher resolution.

"Much work has been done in the last decades to weaken the reductionist position, but it has not been easy. Stuart Kauffman has a very strong argument against reductionism, though it has been challenged. Kauffman points out that, even though we cannot predict the high-level behavior of a system by means of the laws that rule the low level of elementary particles, we can make some predictions using what he calls the laws of complexity. These are laws available at the high level that have nothing to do with the low level. Moreover, according to Kauffman, these high-level laws can be implemented on almost any type of low-level substrate. For instance, they can be implemented on molecules or on a computer, and the laws of complexity are exactly the same. This is a very strong argument, as it means that something exists on the high level that is definitely not there on the low level. Moreover, you can have many different low levels, and yet have the same behavior at the high level.

"Yet even this argument is undermined by several objections. One is that these laws of complexity only rule classes of behaviors, not individual behaviors, unlike the elementary laws, which can also rule individual behaviors. A second objection is that Kauffman's laws of complexity may well have a limited significance: it is easier to predict behavior by using them than by looking carefully at the individual elements. So his argument is only one of efficiency, and no argument of efficiency can give us an ontological proof.

"These objections against ontological emergence are taken very seriously by philosophers and have produced a whole literature. So let me ask the most gloomy question we can raise at this point: Are the reductionists right after all? I know we don't like this, but at the end of the day, are there only the elementary constituents and their low-level properties? My answer is, of course not. Deconstructing the formal concepts of substance and property in quantum physics (as I did previously in our discussions about matter) is as challenging for the reductionist as it is for the supporter of true emergent properties. The physical process may have no substantial roof of emergent properties. It has no substantial ground of elementary processes either. The whole process, actually, is groundless throughout.

"This strong statement obviously doesn't mean that nothing exists at all, which would amount to nihilism, but only that the overall process in which we participate by our actions and cognitive relations has no fundamental level on which everything else rests. It has no fundamental level and no absolute emergent level either; rather, it has a core emergent order. Wittgenstein offers a very beautiful metaphor in his book *On Certainty*: 'One might almost say that these foundation-walls are carried by the whole house.'[16] It's the whole process that is the foundation.

"If the whole process is groundless throughout, then there may be emergence without intrinsic emergent properties—not an asymmetric emergence of high-level features out of basic features, but a symmetric co-emergence of microscopic, low-level phenomena and high-level phenomena. One could call this co-relative emergence of phenomena.

"It becomes clear that the difficulties and paradoxes of emergence arise from a desperate attempt at taking a wrong conception of the low level as a model for the high level. Since the low level is supposedly made of substances with properties, we want the high level also to be made of substances and properties. But it doesn't work. Similarly, if the low-level properties are ascribed causal powers of their own in the very strong sense of productive causality, then one is inclined to ask also for causal powers in emergent properties. So here again, we take the low level as the model for the high level. And this does not work either.

"At this point I am sure that Your Holiness and our Buddhist scholars are completely protected against such temptations to substantialize everything. The Madhyamaka masters criticized relentlessly any reification in the form of belief in intrinsically existing substances and properties; but they also target 'causal power' among the reifications they most strongly dismiss. In the well-known first chapter of the *Mulamadhyamakakarika*, Nagarjuna thoroughly criticizes the idea of properties having causal power as part of their essence. I think it's exactly in this constructive way that we can understand the process of emergence most accurately.

"Let me now consider the crucial case of downward causation from the emergent level to the basic level: for instance, causation from the mental level to the biochemical level. The substantialist framework would demand nothing less than that emergent properties exercise a productive causal power over basic properties. Of course, nothing of this sort is in sight.

"But if you want to think of causal powers in a non substantialist framework, it's very easy. You don't have to think of properties as really producing the events below, such as a real mental property that causes a real physical property in the brain. You just have to ask whether you can act on the higher level. If an event then follows at the lower level—an event that is another type of action—you can say there is downward causation. This downward causation is no longer a substantial type of causality but rather a pragmatic causality."

Michel had touched on a vital intersection between Buddhism and the philosophy of science. Another way of looking at this co-relative emergence is, as J. Schröder puts it, "the influence the relatedness of the parts of a system has on the behavior of the parts."[17] Examples from social science make this clear without resorting to mathematical abstraction: the very constitution of a family, for instance, modifies the behavior of its members. In other words, simultaneously with the emergence of the novel property at the higher level, there is a modification of the properties of the components. The term "simultaneous" is important in the literature of emergence, as it suggests that there is no time shift between upward and downward causality. Indeed, upward causation, or emergence, can be combined with downward causation in the notion of cyclic causality.[18]

The Buddhist Perspective on Origins: Beginninglessness

It fell to Matthieu Ricard now to throw some light for us on the Buddhist understanding of the origins of matter, life, and consciousness. Like so much of Buddhist philosophy, his explanation was couched in a rigorously logical analysis.

"We have heard a wonderful and fascinating story of the origin of the universe and different theories on the arising of consciousness," Matthieu began. "There is a linear gradation, beginning with a kind of extraordinary primordial fire, and then the slow formation and aggregation of heavy molecules. Then matter appears in different ways and allows for more complex molecules, leading to life, and then life becomes more and more complex, leading up to sentient beings.

"If what we have been looking at is the complete story in time and space, then that vision is a perfect description. It will be refined and made more precise, but the basic story is there. But if that story is an episode, a chunk of time and space, then the complete picture could be quite different. The story of science is based on the notion of a beginning, and this is assumed by our basic theory of the universe as we know it. Of course, in Western religions also, we mostly find cosmologies that speak of a beginning.

"The Buddhist perspective puts that notion of a beginning into question, and in a very logical way. A true beginning implies that nothing becomes something; otherwise we're not speaking of a beginning. So how can nothing become something? The Buddhist literature says that a billion causes cannot make something that doesn't exist come into existence. The reason for this is that the supposed quality of nothingness would have to disappear if it were to become something. Nothingness has to shed or get rid of its quality of nothingness to become something, which is impossible, because nothingness has no existence whatsoever. It is only conceived as a mental idea as opposed to existence. You cannot get rid of something that is purely a concept. Changing from nothing to something cannot happen; so observable phenomena, the form of our existence, cannot have come from nothing.

"What could cause a beginning to ever happen? A beginning definitely assumes a first cause, whatever that cause might be. So we need to examine what that cause could be. Is it a permanent cause? Is it its own cause? Does it come from something else again, in which case it is not the first cause? If that very initial cause—and there has to be one—were its own cause, that also doesn't work. Something that is its own cause is already there. It doesn't need to be created. Also, if something is its own cause, that means it doesn't rely on anything else. It has everything included in itself, all the causes and conditions for the next step it's going to produce. The law of causality says that if something does not happen, then the causes or conditions for it to happen are missing. When all the causes and conditions are there, it cannot help but happen. It has to happen. So something that is its own cause would have to produce the same thing forever, constantly and permanently, because as the first cause it is not lacking anything. So that also doesn't work.

"A permanent and immutable cause cannot give rise to something transient. That cause will not be immutable and unchanging because the process of creation is happening, which inevitably modifies it before and after the creation, or the beginning. Causelessness—something happening without cause—would mean anything could happen as a result of anything. Obscurity could come from life; life could come from obscurity. If there is no cause, then there is no law of causality, and there is no reason absolutely anything—a flower growing in the sky—could not happen.

"So you have a situation where, however you examine the notion of first cause, you bump into many kinds of illogical, unacceptable situations, whether it is nothing becoming something, or something that is its own cause causing something else. There are a number of great difficulties in accepting the notion of first cause.

"What other solution is there? Beginninglessness. What is the problem with beginninglessness? It's purely mental. We feel that we can accept going back fifteen billion years to find a beginning, because there's a story for that. But the story should begin sometime; we cannot just go back in time forever. The Western philosopher Bertrand Russell said, 'There's no reason to suppose that the world has an origin at all. The idea that things must have a beginning is due to the poverty of our imagination.' Basically we find it very difficult to think of beginninglessness. But in fact, from the logical point of view, that's the only thing that stands up to analysis. Every other possibility has deep flaws.

"If we accept beginninglessness, of course we still have local stories. We have the big bang and the evolution from matter to life. But that's such a small piece of the story. Given the nature of beginninglessness, we should not try to introduce a new kind of beginning within it, saying that matter came before consciousness, or consciousness before matter. All aspects of the phenomenal world have to be beginningless. Otherwise we are again trying to bring small beginnings into the vast beginninglessness.

"Buddhist cosmology never spoke of a small, limited universe at the center of which we stand. The Buddha spoke of billionfold universes. He spoke of universes that were like curtains of light, like winds of fire, like horses' mouths spouting fire, swallowing fire, all the similes we find beautiful in the photographs from the Hubble telescope. Buddhist cosmology has a very vast vision of time and space. If life, matter, and consciousness don't have a beginning, they are coexistent since forever, in different ways, with different histories, with episodes. There are some local stories where you don't find life, like during the heat of the big bang. But in the vastness of time and space, there is no reason to exclude consciousness and life somewhere else or at other times."

At this point, Matthieu needed to take a breath. He looked around seemingly satisfied at his exposé, and was now ready to make another big leap forward. In his scenario, the big bang and the origin of life, quantum physics, Western and Eastern philosophy, fit together like the pieces of a puzzle.

Touching on Consciousness

"Interdependence means that the different qualities of the phenomenal world are causing each other mutually, or relating to each other mutually, and nothing can be independent. Nothing can really start as an independently existing entity on its own. It can only start from an incredibly complex relationship with all phenomena of consciousness and all inanimate phenomena. That set of relations appears in a particular way depending on how we look at it, or how any kind of sentient being with any type of consciousness looks at it. That's also why quantum physics defines the phenomenal world as observable phenomena. That's what Schrödinger was saying about the world: it is only things that we can experience, that we can observe, and the intimate relation between consciousness and what we observe.

"In that sense, we can say that the state of our consciousness at present is what defines our world—what we can or cannot perceive, what we can or cannot understand. The change of consciousness at different levels, from lesser degrees of intelligence or cognition to higher intelligence—all that was mentioned this morning about Francisco Varela's idea of each one acting on the other—is like two knives sharpening each other. From the Buddhist perspective, the nature of consciousness fashions the world over time, and the way we perceive the world now is the result of the accumulation of experience that consciousness has had, the tendencies accumulated through many lifetimes. That of course is a very strict Buddhist perspective. We call the way we perceive the world now a collective *karma*, a reflection of the whole history of our consciousness that makes us perceive the world in a certain way. The world of a human being will not be the same as the world of a bat or an ant, and it might be totally different from the worlds of other sentient beings that we can't imagine, and that might not be on Earth.

"The notion of mutual causality gives us a different perspective in Buddhism, and I think it is useful in thinking of emergence, of upward and downward causality, and of consciousness. It also resolves the Cartesian dualism that has dominated Western thinking about the relation between matter and consciousness—between matter as a truly existing, solid entity and consciousness, which is supposed to be some immaterial entity. There is an irreconcilable duality and no interface between those two.

"There are two ways to resolve that duality. One is the view that consciousness is just a property of matter. It's the increased complexity of matter that leads to phenomena, or epiphenomena, or emergent phenomena, however you want to call it, which is consciousness. Buddhists resolve the duality by saying that they are of the same nature, but not intrinsically real. We would say that they share the same unreality. We are not going to investigate the nature of reality according to Buddhist philosophy here, but just to say that, in essence, all phenomena, whether conscious or unconscious, don't exist on their own as phenomena gifted with intrinsic properties independent of all other phenomena. In that sense, they are devoid of intrinsic, autonomous, permanent existence.

Consciousness and inanimate phenomena share a common unreality from that perspective."

Matthieu Ricard looked at his watch by rapidly hitching up his monk's robe. He decided that he had still a couple of minutes of the time allotted to him by Arthur, and added an interesting point about the anthropic principle and how it connects with his notion of beginninglessness. I had mentioned the anthropic principle rather critically in my own exposition, dubbing the adherents of this cultural movement "crypto-creationists," but Matthieu had another angle on that.

"There is one other paradox. Certain physicists who hold religious beliefs feel that, instead of having a creator, the universe was so perfectly attuned for life to appear, and for consciousness to witness the beauty of the universe, that some kind of organizing principle—the so-called anthropic principle—must have been there at the beginning to fine-tune the constants of the universe. It's quite amazing that there are a number of constants, some fifteen of them, that deal with the initial conditions of the universe, such as its initial density, the speed of light, and the mass of some basic particles. If these constants had been just a tiny bit different—if the initial density of the universe had been changed by one number after sixty zeros—then matter would not have aggregated in the same way and somehow we wouldn't have life in the end. The precision of that tuning is like shooting an arrow at an orange fifteen billion light years away. Of course it's startling for the mind, and it's difficult not to think that these conditions were made in order for life to appear.

"There are many answers to that view. There may have been many, many big bangs and many universes, most of them sterile, and one out of hundreds of billions gave birth to life. Or there may be many parallel universes. But what I find quite beautiful about beginninglessness is that things have been together from the beginning. Of course they fit together. Is it surprising that the universe is fine-tuned for life to appear, if you think of the continuity? If there are multiple big bangs, then of course there has to be a causal continuity somehow between the former one and the present one. Otherwise we go back to nothing becoming something. That continuity ensures that things belonging to some whole cannot be mutually exclusive. They have to fit more precisely than one in a billion times. The surprise of seeing the fine-tuning is just like putting two halves of a nut together and saying, 'Look how nicely they fit together. They have been part of a whole.' So beginninglessness also resolves the philosophical question about the fine-tuning of the universe, and it makes the anthropic principle unnecessary."

Following Matthieu's presentation there was a long silence. It was that particular flavor of intense silence that one experiences at times in a conference after a major stone has been launched into the pond. I myself was lost in thought on all the questions I would have liked to ask Matthieu. I decided to save some of these questions for an interview with him. The others, as I noticed, were also thinking. Arthur finally moved in his chair, looked at His Holiness, and then opened the discussion, inviting Michel Bitbol to speak first.

A Brief Discussion

"I would like to comment from a Western philosophical point of view on what Matthieu just said," Michel began. "I think that Kant expressed perfectly how Western philosophy and science in general have overcome, or have tried to overcome, the fear of beginninglessness. Western philosophy uses three methods for that, which, according to Kant, rely on three unwarranted reifications.

"The first reification is the reification of the self out of the ceaseless flux of thoughts. The *function* of synthesis of thoughts is mistakenly replaced by a *substance* (the soul, or the Cartesian *res cogitans*) imagined to be bearer of these thoughts. The second reification is the cosmos, which is imagined to be the single hidden cause behind phenomena. Here Kant remarks that science and Western metaphysics go beyond experience into something that cannot actually be checked: we experience a ceaseless series of phenomena, not some exhaustive being like the cosmos that is supposed to encompass everything at once. The third reifying method is God, which is put forward to avoid an endless series of causes. Just as, in a court of law, you don't blame the many causes, beginning with a man's upbringing, that led him to commit a crime, but instead you want to put an end to the series of causes in order to hold the guilty man responsible—God plays exactly this role. His role is to stop the series of secondary causes and to hold responsibility for everything. But perhaps if we didn't fear beginninglessness, if we were more open, we would not need these three reifications at all. This is what Buddhism reminds us."

Eventually the discussion came around to my own presentation. "There was a question that is, in some ways, quite exciting, and we'll see where we take it," Arthur said. "You may remember that Luigi showed us two lists of different categories: one was a list of living things and the other list contained inanimate or nonliving entities. We have a certain way of thinking about the difference, as Luigi said, between these two categories of entities. Now, instead of imagining a Green Man, we could imagine a handsome Buddhist practitioner who is addressing the same question. Would he come up with the same set of distinctions that Luigi made? And if so, on what basis would he make the distinctions?"

His Holiness responded with a complementary question: "Luigi, you've had many, many conversations with Francisco Varela, and I think you understand his view quite well. From my conversations with him, my understanding of Varela's view is that although an amoeba, a single-celled organism with no nervous system, has life and may have some type of sensitivity or reactivity, it has no consciousness, no experience of pleasure and pain. It would have to be a much more complex organism than an amoeba to [have that experience], maybe a hydra. And certainly a mushroom does not have conscious experience of its environment. Is that your understanding also of Varela's view, that an amoeba does not have conscious experience, although it does, of course, interact with its environment?"

I wasn't sure where this was leading. "An amoeba has cognition in Varela's terminology," I answered. "'Cognition' simply means that an amoeba is able to recognize sugar and to repel an organic dye. It does not have consciousness, when this is meant as the conscious knowledge of being able to make such a discrimination."

"At what point do you think that a simple organism is a sentient being?" His Holiness persisted. "At what point does it experience pleasure and pain?"

I responded, "In order to experience pleasure and pain, the organism has to have a much more complex nervous system. If the neural net is too small, it does not arrive at such qualities as feelings."

"But you touch it and it recoils," Ursula Goodenough interjected. "In Varela's terminology, would that be sentient?"

"I think it would not. It would be part of its self-maintenance and unconscious defense, just as it would run away from a poison."

"So avoidance is different from an experience?" Ursula asked.

"Yes," I affirmed.

His Holiness then pointed out a radical difference between the Buddhist and scientific views: "The Buddhist approach is to distinguish generally between sentient and nonsentient beings. There is no distinction made systematically, in the scientific sense, between nonliving and living organisms." This theme would come up again in our later discussion of ethics, but for now there seemed to be a willingness to reconsider the Buddhist position.

"This is a point where Buddhism needs some editing, some modernization," Matthieu Ricard offered.

"You could be our editors," said Alan Wallace, stepping out of his translator's role for a moment.

"We are learning," His Holiness said, nodding seriously.

But the scientists were no less eager to learn, and Ursula pressed on with a question that she hoped might "begin to raise some larger questions that will come later on. For example," she said, "a key element in living systems, in plants and certainly in living animals, that is not associated with sentience is development. They may start very simply and then become increasingly elaborate in their structure. One can think, for example, of an embryo, whether a chick or a human embryo. It begins very, very simply, and there is no question of sentience then. It's one or two cells at first, and then it begins to develop more and more complexity. At some point, it may be able to feel pleasure and pain, but not in the beginning. This increase in complexity through development is a very powerful and important idea and needs to be explained and accounted for in Western science. Can you say anything from the Buddhist standpoint concerning development? Are there energies present at the beginning? We have heard about *pranas*, for example. In the West, in the eighteenth and nineteenth centuries we had an idea of some kind of vital energy, which we have now abandoned, by and large. Can you tell us something about the role in development of pranas or energies that are associated with living systems?"

The Dalai Lama responded, "We have a kind of parallel explanation of development that gives an account of how different types of prana energies emerge at different levels of the physical development of the embryo. There is also a basic understanding that, in fact, the very function of directing consciousness toward an object is thought to be a function of this prana energy. The prana energy is thought to be behind many activities that have a function of directing something. But I don't know much about the details of these theories."

Reflections

As we started moving toward the exit at the end of the session, most of us were absorbed in thought. My own mind was still on Matthieu's presentation about beginninglessness. At a recent conference I had heard from John Barrow about his concept of the multiverse, a theory that, coming from as far away as string theory, foresees a series of universes originating one from the other,[19] an idea put forward also by the astrophysicist Leonard Süsskind.[20] These modern theories, which also leave behind the idea of a beginning of the universe, have developed without any link to the Buddhist view of beginninglessness. It is interesting how similar things come from different sources. I was also thinking about the relationship between Matthieu's notion of beginninglessness and my own talk about the origin of life, the scientific view that foresees a neat beginning of this event. In principle, there was no contradiction between the two, as my scientific notion clearly concerns a "local story," the story of our planet Earth, a small speck of dust in the universe. Is there no contradiction then between science and Buddhism in this respect? Not quite. My scientific theory of the origin of life implied that life, sentience, and consciousness would derive from the same material basis—that once the structure and self-organization of the cell has appeared, life would originate as an emergent property of this material basis. Moving from the simple first cells to multicellular and eventually to higher organisms, cognition and perception would arise as the sensorium of the organism became more and more sophisticated. At a certain point it would acquire sentience and consciousness. All this comes from within the structure, with no place for any transcendental element. This is not what Buddhism teaches us. We will see more of this contrast in chapter 6, expressly devoted to consciousness.

4 / How Life Unfolds

AND AN INTERVIEW WITH RICHARD GERE

WE HAVE FOLLOWED HOW science describes the transition from inanimate matter to life as an allegedly spontaneous and extremely lengthy increase of molecular complexity that eventually reached the point where the first family of cells formed. These pristine cells were capable of self-reproduction and could therefore propagate and multiply. This is the point where the discussion on the origin of life ended, giving place to the next conversation, on cellular evolution. Arthur Zajonc opened the day's meeting, summarizing the ground we had covered thus far but also commenting on what he perceived as "the joy of being a Buddhist intellectual."

"In the West," Arthur explained to His Holiness, "sometimes intellectual life is earnest and serious, but the mood of our conversation here, in this week, was also filled with a pleasure, a joy, a happiness of the sort that you talked about in your opening remarks. There was also a trust and reliance on reason and logic, which became so evident during the course of your presentation as well as from the scientists' side; and then also the valuing of evidence, the constant search for an empirical basis in our discussions. These elements played into what I would call a skeptical mind, with questions such as: How do we know these things? What can we know to be true?"

Arthur was representing all of us when he expressed his gratitude for the remarkable tenor of the discussions. And then it was time for Ursula Goodenough to begin the morning's presentation, which would take us into the mechanisms of a cell's life and the foundations of biological evolution.

Ursula Goodenough presenting the cell and evolution between Matthieu Ricard (left) and Michel Bitbol (right).

The Cell Makes Proteins. . . .

"I would like to talk about the evolution of life," Ursula began. "I will start by talking about how life works at the molecular level, and then, using that understanding of life, I will talk about how it evolves. So let's start with the cell. All of the modern creatures on the planet are either single-celled organisms or organisms that are made up of cells. The cells of all creatures are all organized in the same way. As we have seen earlier, each cell has a membrane around it that separates the outside from the inside. Inside, there are enzymes that mediate the metabolism, the breaking down of food to make energy, and the biosynthesis, the making of big molecules from small ones. In addition, the cell needs to moves things between the outside and inside; that's done by proteins called channels, which transport molecules in and out of the cell in a very regulated way. Finally, the cell needs to know where things are and what's happening around it; that's done by another group of proteins known as receptors, which mediate interactions with the environment.

"The cell knows how to make all of these things. Its genome has the instructions for making that kind of cell, and this genome is copied by the cell and transmitted to its progeny. Each genome is a collection of genes—all of the genes that a particular kind of creature needs to instruct itself. A bacterium has about 5,000 different genes, an insect has 20,000, a plant 25,000, a human 30,000. Each one of these genes encodes the structure of a protein. A bacterium, we would expect, has the capacity to make 5,000 different kinds of proteins."

The Dalai Lama was intrigued. "Are these 5,000 different types of genes? And each of the different 5,000 genes can replicate themselves?" Ursula affirmed this, and he continued, "How long will a gene survive? What is the maximum lifetime of a gene, maintaining its identity?"

"Genes are very stable," Ursula responded. "These DNA molecules, being wrapped around themselves, are very stable." A slide showing the structure of DNA illustrated her point and elicited much interest, particularly from the Buddhist group.

"Is that actually a photograph?" the Dalai Lama asked.

"Yes," Ursula replied. "That is a real piece of DNA. It's shadowed with metal particles and observed by the electron microscope, magnified maybe 500,000 times. If we could unravel it, we would see that there are, in fact, two strands, the famous double helix of DNA, and if we unraveled it even more, we would see that in each one of these strands there are residues of individual molecules called nucleotides." She identified the separate nucleotides: adenine, guanine, thiamine, and cytosine, abbreviated A, G, T, C.

"The order of the nucleotide sequence is what's important," she continued. "It is the order of the nucleotides that tells the cell what kind of pro-

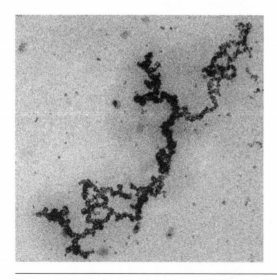

Electron microscopy of DNA, made visible by absorbed copper sulphide nanoparticles.

tein to make, and it's interpreted as a code. It is like the translation of an English word into a Tibetan word. The apparatus, or factory, that reads and translates the nucleotide sequence into proteins is a large structure called a ribosome. So, for example, if it sees the nucleotide sequence ATG, it knows that this word stands for an amino acid called methionine. The next triplet might correspond to an amino acid called alanine, the next to an amino acid called histidine, and so on. . . ."

The Dalai Lama asked whether the nucleotides could only be read in clusters of three, which Ursula affirmed, and why that was so. "I would love to spend the morning explaining this to you," she said, giving voice to a feeling that every scientist who presents at the Mind and Life conferences has experienced when facing an eager and extraordinarily intelligent listener under the pressure of limited time. There was a vast amount of territory still to cover, and here Ursula had to speed through the difficult biochemistry of protein synthesis in the ribosome: how a gene from the DNA genome is copied in the form of messenger-RNA, and this single-stranded molecule can leave the compartment of the genome and migrate onto the ribosome; how different molecules of transfer-RNA, each carrying an amino acid that matches a specific triplet, bind in turn momentarily to the appropriate triplet on the messenger-RNA, and in doing so, donate their amino acid to a growing chain that will become a complete protein.

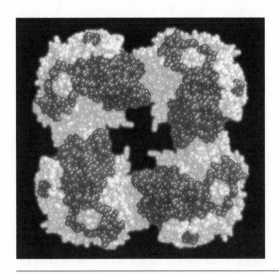

The three-dimensional molecular model of the structure of a complex protein (the transacetylase core of the pyruvate dehydrogenase complex, which consists of twenty-four identical chains; twelve can be seen in this view).

"Once this string of amino acids is made," Ursula continued, "these amino acids start to bind to each other. The chain folds so we get a protein that has a shape."

"Are the shapes definite?" asked His Holiness.

"No, the shapes expand and contract, and change. But at any one moment, in a population of the same kind of protein, most of them will adopt a similar shape. The shape is what they use to do things: like a gear in a watch, the protein's shape is important in how it fits into the rest of the machine. Many structures in the cell are made of groups of proteins whose outside surfaces fit together, forming little machines or apparatuses. Also, it's these proteins' shapes that will change when cells evolve; they will make different shapes and then be able to do different things.

"When a protein is carrying out its biological function, it will change its shape." Ursula illustrated her point with a cartoon of an enzyme involved in catalyzing the formation of a chemical bond between two sugar molecules. "The enzyme has two pockets into which the sugar molecules fit. Once the pockets are full, the enzyme changes its shape. It's that shape change that gives the energy and positions the sugar molecules correctly so that this chemical bond can form."

The Dalai Lama asked how long the molecules remained in this state and seemed surprised at Ursula's answer: a microsecond! "Once the chemical bond is formed, it no longer has the right shape. It no longer has the affinity to stay in there. It pops out and two more come in."

Receptors

Ursula went on to introduce one of the central issues in the molecular biology of proteins with a beautiful image of a receptor protein protruding out from the cell membrane as well as into the cell.

"This is a protein that serves as a receptor. Suppose now that there is a stimulus, for example an odor, a molecule that emits a smell. The receptor has the right shape for that odor molecule. In our nose we have about two thousand different receptors, all subtly shaped to recognize a particular odor. Once this odor molecule goes into the receptor, it changes its shape. This change affects the entire shape of the molecule, and produces a chain of successive changes as the cell responds to the odor.

"When you smell, you detect the smell within less than one second. All of these shape changes are happening very quickly, through the brain and into the mind. When the odor molecule leaves the pocket, the receptor goes back to its original shape and another molecule may come in."

The subject of odor precipitated a series of questions from the Dalai Lama on sensory perception. It is a subject that has long fascinated him and come up at several previous Mind and Life conferences, as it is an area where Buddhism also has elaborated precise theories. Ursula explained that hearing, vision, taste, smell, and tactile sensations are all determined by receptor proteins of one kind or another. Whether the stimulus is a vibration or a chemical, it changes the shape of the receptors, and that change triggers a chain of other changes that reach the brain. But questions remained.

"We have five types of sensory stimuli coming in," His Holiness began. "And we have five types of physical sensory faculties. Right? So which comes first? Are these sensory objects, the stimuli, already out there and simply being detected, in which case there could be a lot of other sensory objects out there that don't correspond to these five faculties? There could be hundreds of them. Or are these five types of stimuli actually produced by the faculties, in which case there are five, because we have five?"

Ursula was taken back, but replied almost immediately, "Light, for example, has always been out there, and the organisms have evolved receptors to detect it. It is very important for a plant to have light, and we have receptors for light so that we can find our way in the world when we move."

"But the crucial point here," His Holiness continued, "is that, although there are photons of many frequencies, none of the photons or electromagnetic fields is blue. There are no photons with colors, no electromagnetic fields with colors. And so, although they are out there before anyone has any sensory faculties, there's no blue out there without sensory faculties. I'm asking about the appearance of light, and not the objective, physical substrate of light."

"The blueness is done by the brain," Ursula responded.

"Ah, so it's actually created by the sensory faculty."

"Right. The receptors in the retina of the eye are very sensitive to particular wavelengths. What we call red is a 670-nanometer wavelength of light, and specific receptors change their shape when that kind of light comes in. We then interpret that as red."

"So in fact, the molecules that align themselves in the cells in the nose are not odor molecules, they're just molecules," the Dalai Lama concluded, restating the case in a way that aligned more closely with Buddhist theory. "The experience of smell is actually generated by the faculty. There are no taste molecules. The experience of taste is generated by the tongue faculty, and so for heat and cold, smooth and rough, all of these are generated by the faculty, stimulated by objective molecules and frequencies of energy. These are not actually out there independent of the faculties."

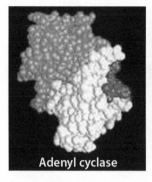 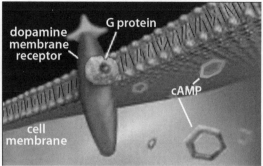

Examples of receptor proteins (adenylyl ciclase and dopamine) in simple schematizations.

Ursula wrapped up this section of her presentation by showing a series of actual microphotographs of different types of receptors. Their convoluted complexity made a dazzling contrast with the cartoon versions of the same molecules that had illustrated her story. "All the shapes of these receptors are ultimately encoded in the gene," she emphasized. "The gene has the information to make those shapes. . . . Now you know almost everything you need to know," she concluded, if a little prematurely.

The Dalai Lama still had questions. "Are these shapes of different receptors the same for human beings and animals?"

"Yes, they are very similar in most cases," Ursula responded.

"And there is no difference in age, for children and grown-ups?"

"There are some receptors that will appear in adolescence. When a person becomes sexually mature, then new receptors get expressed. That takes me to my next point: these genes do not make their proteins all the time. There are switches: some of the genes are turned on and expressing their proteins; other genes are turned off and not expressing their proteins. Genes for making sex hormones just sit there until adolescence, and then they're switched on. If we look at a bacterial cell, even though it has 5,000 genes, maybe only 3,000 of them are turned on at any one time, and the others are not being expressed. The gene expression—its translation into a protein—is regulated.

"At the beginning of the gene, there is additional DNA that works as the light switch. There are lots of names for it, but we'll call it a 'promoter,' because it has the capacity to promote or not promote the expression of the gene. This process is in turn regulated by another protein, called an 'activator,' which recognizes the nucleotides of the promoter and permits the whole apparatus to read out the protein. The gene expression, as we say, is switched on.

"There are other proteins called 'repressors,' which have the opposite effect on the promoter, switching it off. They keep the gene from being expressed. If the repressor goes away and the activator goes on, the gene can be switched on again. So what cells do then is decide. They're intelligent. . . ." Ursula was speaking metaphorically, of course, and did not mean to ascribe any consciousness or volitional choice on the part of the cell, but the translators had to untangle the unintended implications. "In this way, cells decide on the basis of what they need—what they perceive from the outside—what repressors and activators they produce, which in turn determines which genes are expressed.

"An amoeba has maybe eight or 10,000 genes, and it decides which genes to turn on depending on what information it is receiving from the outside. But all this is much more important in multicellular organisms, like us. We're a whole organism, but we're made up of some trillion cells. As you know, these cells are different from one another. We have skin cells, liver cells, and so on. Every one of the cells in our body has the same genome, the same 30,000 genes. What determines whether you get a liver cell or a skin cell is the following: some of the cells in your body switch on genes that say 'Be a liver cell.' Others switch on genes that say 'Be a skin cell.' It is all done by this switching on and switching off." Again I noticed difficulty in the Buddhist camp with these expressions, as if the cells or genes could "think" and "say." "No, no, genes and cells do not have consciousness," pacified Alan and Thupten in Tibetan. "It is just a way of speaking."

From Genes to Embryo

There was a brief pause. Having made her points about receptors and the regulation of cell activity, Ursula was ready to switch gears. She looked around to gauge the comprehension of her audience, and started again with energy. "I think you're familiar with the way that multicellular organisms begin with a sperm and an egg, which fuse together to form what's called a zygote. The cell then copies its genome and divides in two, but the two cells don't go away from each other the way they would if this were a bacterium or an amoeba. They stay together. Each of those two cells, then, copies its genome and divides again, and we have four. But they also stay together. And then we have eight, and then sixteen. But in a particular cell, a gene may be switched on to make a protein that's not in any of the others, and that may cause other genes to switch on. In this way, different cells will make skin, or muscle, or an eye."

At this point, Ursula displayed a slide of an embryo that would become the larva of a fly. "It's divided a lot of times, so there are many cells here already. But only these cells down here at the bottom of the larva," she pointed out, "have switched on a particular protein called a twist. And these few cells, because they have this twist protein, do something new in the embryo. They move inside the embryo, and they will become the body wall of the embryo's gut. A huge number of scientists are studying this switching on and off of proteins, and we really have an elegant picture now of how embryos do this patterning. The same thing happens in adults. In your own red blood cells, hemoglobin is being made. There is no hemoglobin anywhere else in the body: it is switched off except in your red blood cells, where it's switched on."

Genes in Action

Until now, everything Ursula had been explaining related to the biological cell. This was all preliminary to the life of the cell as an entire organism and to the life of a colony of cells as a community. After our well-earned tea break, she was ready now to turn her attention to the bigger picture, but those protein shapes and their changes would not be forgotten, for they form the basis of most larger-scale actions in biology.

Ursula looked at His Holiness, who as usual did not seem tired at all, and then at her colleagues, who also seemed eager for more. She started again: "Now that you understand life at a molecular level, evolution is quite easy. We're going to be looking at changes in the genes, leading to changes in the protein shapes, leading to changes in the emergent properties. Once you understand how life works, evolution makes a lot of sense.

"Let me remind you what an organism is, from our present perspective: an organism is the integrated sum of its biological traits. All of these traits are organized and balanced with one another, and these traits are emergent from protein shapes and shape changes, which in turn are encoded in the genes.

"The organism, of course, is not just living in a vacuum. It's living in an environment, as Luigi pointed out for us. Although it's hard to define, its environment has parameters, and we call the collection of environmental parameters a 'niche'—where an organism actually finds itself. Basically, the proposition in life is to negotiate the niche: to live in that context, to find food, find mates, find shelter. Hopefully the instructions that the organism has are a good match for the niche. You could say that the genome is sort of a proposition as to what the environment might be, and the organism will negotiate that environment, copy its genome, and pass it on to its progeny. Through progeny, the genome moves through time.

"There's one more word that we need here, which is 'adaptivity.' We say that an organism is more or less adapted to its niche. If it's well adapted, it flourishes and thrives; it has good health and lots of children. If it's poorly adapted, that means it's not doing so well, and it may not be successful in having its offspring live through time.

"When we talked about genes, we considered them as stable entities, and that is in a way quite true. Genes are very stable, and they are usually transmitted from one generation to the next very accurately. But nothing is perfect, and occasionally when the genes are copied, or when environmental agents such as ultraviolet light come into the cell, mutations can occur, which change the sequence of the nucleotides. A mutation will produce a variation in the protein shape. In the particular case we're going to look at, that leads to a variation of a trait of the organism. Evolution then entails a selection of organisms that are better adapted to their niche.

"As an imaginary example, suppose that an amoeba has a mutation that makes it move differently—better—than other amoebae in the same niche. Moving is very important for an amoeba. For an amoeba, moving and eating are the same thing: it has to crawl toward its food, then eat it.

"Now what's going to happen is selection. I'll give you two scenarios for what might happen to amoebae with different motilities. One possibility is what we call a 'sweep.' If this new gene—the mutation—makes the amoeba have a much better motility, then very rapidly, after a few generations, all of the amoebae in that niche start to carry the new gene, as they will have many more offspring. The amoebae without the new gene can't get the resources as easily, and so their genes fail to spread through the population, and the new gene sweeps the population.

"But in fact there are all sorts of possibilities, and I'll give you one more. Suppose that an amoeba without the mutation moves very well and gets food easily if the soil is wet, and an amoeba with the new gene is better able to crawl when it's dry. Because soil in this niche can be wet or dry, in wet seasons you will find a lot of the amoebae without mutations, and in dry seasons, more of the other type. Looking at the population as a whole, you find both versions of the gene in different amoebae. This kind of diversity in a population is very important. It's important that in a human population we have lots of different versions of the same gene, the same trait, because this allows certain people to do well in certain situations, and others in other situations. The term for this kind of diversity is 'polymorphism,' and it describes a healthy population. A population that's very specialized is very vulnerable."

Here Ursula made another very important point: "These mutations are random, as near as we can tell. Whether a given gene in this huge genome undergoes a mutation, where it occurs, and what the mutation looks like, all happens randomly. But the process of natural selection is not random. It's very choosy. It's going to work or it's not going to work. This is very, very important.

"I bring this up because there's a common misunderstanding that evolution is just a random process, and that is not true. Even though the ideas—the new protein shapes and new traits—are being tossed out randomly, the outcome is very, very dependent on the environment and the adaptivity of the particular result."

This notion is related to what we said in the previous chapter about autopoiesis, and in particular about the relation between autopoiesis and evolution: only those environmental changes that find a correspondence with the internal organization of the autopoietic unit will be accepted.

Ursula paused again briefly. She was now ready to bring her audience to the narrative of evolutionary creation. "Now, in the amoeba story we started out with one kind of motility and moved to another kind of motility. We were changing something that already existed, and that's what usually happens in evolution. But occasionally there is a sequence of events that actually

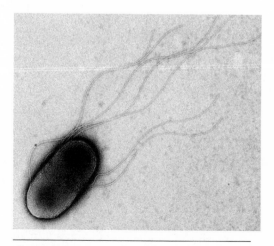

Electron microscopy of a bacterium with flagella.

makes something new. I'm going to tell you now how bacteria acquired motility. We do not actually know whether it really happened this way; however, billions of years ago, something like this must have happened."

The Birth of a Flagellum

"In the beginning, bacteria could not move. They just sat there, but at some point a bacterium developed a flagellum. A flagellum is like a small propeller that makes a little bit of current in the water, and this allows the bacterium to swim. In this way the bacterium can move toward food. The question is, how do you go from a bacterium that has no flagellum to a bacterium that does have a flagellum?

"The nonmotile bacterium has protein channels that allow small molecules or ions to move in and out of the cell. These channels are very important for its survival. We now have a random gene mutation that results in a protein with a shape that just happens to be able to bind to the channel. As a result of this protein binding, the channel works better. And so this trait is selected, since a bacterium that makes this protein is better adapted. But another thing happens when this protein binds: the channel actually spins around in the membrane. We now have a rotating channel, and another random mutation produces a long fibrous protein attached to it; then we get our little propeller.

"It's much more complicated now. The modern bacterium's propeller has about twenty-five different proteins, some of which regulate the speed at which the flagellum goes around or reinforce its attachment to the channel so that it doesn't fall off. Once something works, then selection comes into play to make it better. The real trick in evolution is the first novel moment, the first appearance of something new."

Now came a typical question from His Holiness with a strong Buddhist flavor: "Has the bacteria's propeller been verified by observation, or is it inferred on the basis of its movement?"

"My account of the sequence of events was speculative, but the flagellum itself is very well understood. It has gears so it can go fast, or it can stop, change direction, and start up again. It's very nice."

"Isn't the sperm cell similar?" asked His Holiness.

"The sperm's flagellum is a different thing altogether and much more complicated. Unfortunately the same word is used for both."

Ursula then presented a more complex example of mutation that occurs not in the nucleotide sequences that express proteins, but rather in the segments of the DNA that act as promoters or as activators. "Promoter and activator mutations occur often, as often, in fact, as the mutations in the protein-encoding gene itself. They also have consequences for the organism." Her example showed how an activator that controlled the growth of the legs in insect larvae had mutated in butterflies to allow the growth of legs from the belly as well as the chest. "These mutations in the regulatory systems are particularly powerful in embryology, because the timing and the place where genes are expressed in the embryo has everything to do with how the embryo differentiates."

Ursula concluded this section of her presentation with an important observation about evolution that has only become clear with our understanding of the molecular processes involved: "We've seen how diversity arises in the example of the bacterium getting the flagellum. All of these new developments generate organisms that are adapted to new niches, or better adapted to old ones. However, we also see a great deal of conservation of ideas. If you have a protein shape that fulfills a very important biological role, then the gene encoding for that shape tends to continue through time. There is a great deal of conservation of structural genetic information, as well as new ideas, moving through time. When we think of evolution, it's very important to see both."

The Intelligence of Hox Genes

Ursula looked at the clock. There was time for a last item, one that involved yet another jump of complexity. Her theme had to do with the question of how an embryo makes its body plan—how the head and the body are differentiated, and how vertebrates get their arms and legs. The key here is hox genes, the so-called molecular architects. In 1978 an American geneticist, E. B. Lewis, published an extraordinary report on how mutations in the fruit fly *Drosophila* produced gross alterations of the fly's body plan, with legs in place of antennae. These monstrous fruit flies had the honor of shocking the mass media and were instrumental in the proposal by Dr. Lewis of a remarkable concept: that these developmental errors were caused by a mix-up in the functioning of "master control genes" that normally control the development of the fruit fly's body in defined segments. Workers in the field quickly set up experiments to study and isolate these genes indicated by Lewis, and eight master regulators were discovered. This was followed by an even more remarkable discovery: that the same eight genes, with only minor variations in the genetic code, were present all across the animal world. These became known as "homebox" or "hox" genes, and it was subsequently discovered that mammals possess four sets of hox genes, in contrast to the single set controlling development in the fruit fly. In other words, all mammals on Earth share the same master control genes to make their anterior-posterior body plan.

Ursula began by relating the hox genes to the importance of conservation in the process of evolution. "If we compare the genome of a fly and the genome of a human, we find a lot of genes that code for proteins of the same hox family. In the fly, the gene that specifies the head region is called OTD, and in the human it's called OTX. In the fly embryo, all of the cells that will become the anterior part of the fly are initially specified by this gene, and the parts of the embryo that will become the more posterior parts of the fly are initially specified by different members of these hox gene families.

"If we look at a human or a mouse, it follows the same general pattern. The whole idea of setting up an embryo with a head and a middle and an end is conserved in the insect and the human, and it's done with the same family of regulators. In fact, since one can make mutations in the laboratory, one can remove the OTD gene from the fly's genome. If you do that, of course, the fly can't make a head and it dies. But if you take that fly and put the gene from a human into the fly, the fly makes a head. It makes a fly's head, not a human head. So what this protein does is to specify 'headness.' The OTX or OTD gene sets up part of the embryo to become a head, but it activates fly-specific genes in the fly, and human-specific genes in the human."

His Holiness asked the question that was on many listeners' minds: "What if you reverse the process?"

"For ethical reasons this experiment hasn't been done in the human, but it has been done in the mouse, and the answer is that deletion of the OTX gene in the mouse produces headless embryos, and insertion of the OTD gene from the fly allows such mice to form heads—again, mouse heads, not fly heads."

"So there's no real genetic basis, as such, for different parts of the body?" the Dalai Lama wanted to confirm. "It's more the switching on and off of the regulators?"

"Yes," said Ursula. "Embryology is the same proposition everywhere. The principle is always the same: once a gene is turned on, it has a result that causes another result that has another result. In the case of the anterior part of these two kinds of embryos, if this OTX gene is there, then it switches on certain factors that determine the head direction. That's also true in the nervous system."

Having covered so much ground, Ursula was ready to embark on a discussion of macroscopic evolution. She took a big breath. "In evolution, we

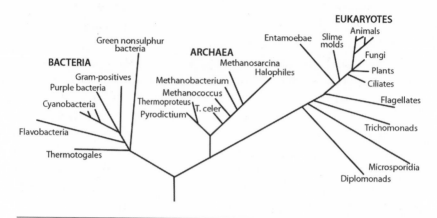

The three main initial kingdoms of organisms on our planet (archaea, bacteria, and eukaryotes) and their later evolutions.

started with the fish with no limbs. The first sign in the fossil record about fifty million years later is a wonderful creature that has things that look like the fins of a modern fish. But that's not what they are. They are the precursors for arms and legs.

"In the next creature, about a hundred million years later, these fleshy fins started to become flippers, which look similar to our arm, with a humerus, a tibia, a radius, metatarsals, and digits. These first limbs were not developed on land. They were actually used to swim. However, about ten million years later, we start finding this invention on land. It is again a situation where we have the result of an unintended consequence. These appendages could move in the water, but all of a sudden the creatures realized that they could crawl up on land. The plants had already come to the land by that time, so there was lots to eat. So it was a really great idea to use those things on land. So again we have novelty, something new.

"And again, it is hox genes that are switching on to make these limbs. In fact, there are two particular hox genes that get switched on in the anterior part of the body to produce these limbs. This is co-option: existing hox genes are being expressed to generate a new idea.

"Finally, once a good idea has been introduced, then you can play with it. For example, very long fingers developed from those appendages in different vertebrates such as the bat or the human. Alternatively, the digits disappeared, as in a horse. But it all diverges from the basic idea of making a limb, adjusting to get adaptive kinds of limbs for different niches.

"To conclude, we return to our original cell with its receptors, its enzymes, its membrane, its channels. We understand that this cell has a genome with lots of genes that are coding for proteins, channels, receptors, and so on. If you look at a given kind of creature, you may find a very new set of genes that's specific for that particular organism, but we also find that a lot of genes are conserved, like the hox genes or the genes for certain receptors. They are conserved as good ideas that move through time.

"Once we find one of these conserved genes, we can sequence the same gene in a plant, an animal, an amoeba, and, in some cases, in a bacterium. What we conclude from this is that all the creatures on the planet have a common ancestor, because the same genes are found in all modern creatures.

"In fact, we can start to construct evolutionary trees." Here Ursula displayed several versions of evolutionary trees, based on analysis of commonly shared DNA.

Three Directions of Life

"The first insight is that life has gone off from a common ancestor in three directions. One is the bacteria, which are very familiar to us. The second group is less familiar, called the 'archaea,' which tend to live in very extreme conditions such as hot springs and geysers. The third group are the eukaryotes. These are the ones that we are inevitably the most interested in because we belong to this group."

"Has the differentiation between plant and animal already happened at this point?" asked His Holiness.

"No, the plants are all eukaryotes, as are animals and fungi."

Ursula explained her diagram, where the length of the lines shows how closely related the genes are to one another. Although they don't really show time, the ratio of the genes is for all practical purposes a molecular clock showing how recently the different organisms departed from a common ancestor.

"All of these early organisms that come from the first ancestor are single-celled. A lot of them are amoebae, because the invention of an amoeba—a single-celled organism that crawls in a certain way—happened many times in evolution. Other single-celled lineages come later in time, such as the ciliates that swim in ponds.

"But our anthropocentrism focuses our interest on three groups, the fungi, the animals, and the plants, because they have multicellular components. They're clearly three different ideas, but they all share a common ancestor. The fungi have experimented with being multicellular, and some have remained single-celled. Likewise the plants: some of the plants are single-celled, such as green algae, but we also have the trees and the flowers."

Enter the Hominids

"Lastly, we have the animals, and they diverged quite early, some 500 million years ago, into the invertebrates and the vertebrates. All primates have a common ancestor. The monkeys and apes diverged about 60 million years ago. In the great ape lineage, the orangutan went off on its own direction first. Between 5 and 7 million years ago, the hominid lineage took off. There were other apes that were still around, and about 2 million years ago there was another divergence between the common chimpanzee and another chimpanzee family called the bonobo. And there were lots of hominids before we get to the human."

"So in terms of seniority, orangutans and gorillas are senior?" asked His Holiness.

"Well, no," Ursula corrected him. "That's not the way to think about it. These are all modern organisms. The question is when a lineage broke off from a common ancestor and began going in its own way."

The Dalai Lama asked when *Homo sapiens* first occurred. "It depends on how you define it," Ursula responded, "but 100,000 years ago, *Homo sapiens sapiens* was already around. The cave paintings were made 40,000 years ago and agriculture began 10,000 years ago. We are very, very recent."

This brought the morning session to a close, and with it Ursula's overview of a very complex subject. Later in the conference, she had the opportunity to take the hot seat again and continue her presentation, returning

The evolution of hominids, from the very beginning to the bitter end.

to the theme of emergence but focusing now on the emergence of human consciousness.

Genetic Cultural Masking

"In our brains there are about a hundred billion neurons, and they interact with one another to integrate signals and to produce emergent properties that we call our brain-based awareness," Ursula went on. "Brain-based awareness exists in all of the animals and, in particular, quite sophisticated brain-based awareness is found in all of the vertebrates.

"When we come to the primates, we know quite recently there was a divergence of the hominids and an evolutionary sequence leading to the human. We know that we share a common ancestor with these closest relatives of ours, the chimpanzee and the bonobo.

"The question is, what is distinctively human as contrasted with these brother and sister creatures of ours? We can approach this question in two ways. We can talk about mental traits that we share with these other animals, like intelligence. We share a dependency on nurture for our mental development: a baby chimp without a mother cannot thrive. We share similar kinds of temperaments. There are chimpanzees that are very shy and others that are very aggressive, and humans have these same traits. They are emotionally quite similar to us. We all learn by experience and by imitating others, and because we are all social animals, we have social, political, and hierarchical interests in forming social systems. And because we are social animals, we also have a number of pro-social emotions, such as empathy. These animals like to play and laugh, and they care.

"To me, our relationship with these animals is deeply important and very grounding. At the same time, we can also ask what makes us distinctive. For humans it is particularly important that, in addition to imitation and experience, we access information from our language-based cultures.

"Let me first talk about culture. What do we mean by this? Humans, like beavers, are adapted to niches of their own construction. A beaver constructs a dam, and then beavers are selected for their capacity to construct and thrive in dam-created niches. Similarly, humans construct cultures and we are then selected for our capacity to construct and thrive in human-created mental niches.

"So we are responsible for our niche. This generates a very interesting dynamic in that we can talk about the coevolution of culture, language, and brain, or mind. Human culture is encoded in and acquired via these languages. These languages have been selected for their ability to be learnable by children's brains. And similarly—and round and round—children's brains have been selected for their ability to learn symbolic languages. Therefore, the acquisition of language-encoded cultural understandings has coevolved with the emergence of symbolic mind. Something is set up here that can evolve quite rapidly.

"How might we model the evolutionary events that lead to the emergence of human minds? We don't know, but one idea is something called masking. The best way that I know to describe masking is with the example of vitamin C, or ascorbic acid. Almost all organisms on the planet make their own ascorbic acid, which is necessary for their survival. They have genes to do this, and the corresponding enzymes make ascorbic acid. But the apes started eating fruits that are rich in ascorbic acid. As a result, mutations affecting the enzymes involved in the synthesis of ascorbic acid were masked from natural selection, because the animals were getting ascorbic acid from the outside. If you have a mutation that prevents you from making your own ascorbic acid, but you are getting it from the outside, there's no selection for not having a functional version of that gene. Because things tend to degrade, it's very easy for a gene to lose its coding capacity through mutation. But in the case of masking, there is no effect [on coding capacity].

"In theory, cultural masking would lead to the dependence of hominids on culture. In other words, if culture provided hominids with useful information from the outside, any hard-wired genetic programs that specify the same information would be masked from selection. Because the necessary information comes from the outside, the genetic programs would be no longer operant. The notion is that hominids, therefore, became dependent on culture for their survival.

"This is certainly true of the modern human. We need information from our cultures to know how to take care of our children and how to interact with one another. We cannot function without culture. One hypothesis is that the brain programs that were degraded by masking were reconfigured to have language capacity. If you need culture, which implies needing language, then other brain capacities might have been degraded but reconfigured to use language. In this case, degradation would be adaptive because you would be better able to get this information from the outside and to create your own niche. This is testable. In the laboratory we can ask whether ape brains and human brains are different in this respect.

"We have a lot of ideas about how this brain program degradation might happen. We know that our brains have become larger, so having more

Ursula Goodenough discussing genes and evolution.

neurons might help. But we could also rewire the brain capacity that's already there. It turns out that the brain is not very genetically determined. The cells in the embryo that grow into the brain have general instructions for where they're supposed to go, but what happens then is much more of a self-organized process. That means that if you change things quite modestly at the beginning, you can get quite different results. If you just change the starting conditions, you can get quite a different brain.

"So the model is that culture has masked the need for many genetically encoded mental pathways, and they have degraded, and the freed-up brain space—we don't really know what we mean by that—has been reconfigured to generate minds adept at learning symbolic language, and hence acquiring cultural information. What do we mean by this symbolic language? How does it work?

"There are three general kinds of encoding systems that have been proposed: iconic, indexical, and symbolic. An iconic system is basically one for one. An example we've seen is the way that a nucleotide triplet encodes for a particular amino acid. Cellular awareness is also iconic: a hormone binds and the cell responds, one on one. Brain-based awareness uses an indexical system, in that several different things come in together and the brain integrates them and produces an output. In symbolic systems such as language, collections of these indexes are signified by abstractions that we call words. We have a word for a whole nested set of indexes, and we can then manipulate these words and use them all by themselves without having to refer to the lower levels. We can, if we are challenged, unpack them. We can say, 'What do you mean by *information*?' But we have the ability to use the words themselves. We have a whole new virtual reality.

"This symbolic language allows a new and emergent level of semiotic freedom. We are no longer bound by antecedents, in a way. We could make a long list of what this gets us, but I'll just point out two things. One is called conceptual blending: we can put words together and reconceive or transfigure them, moving them around to create new ideas, new whole concepts. We can say something like 'love thine enemy.' An ape cannot do that.

"And somewhere along the line, we also developed the symbolic experience of ourselves. I'm not saying that we *only* experience ourselves symbolically, but we do experience ourselves symbolically. We have a self-concept that we can describe in words, and much of our imaging, our logic, and our conversation has to do with this self-concept, which is, I believe, a linguistic concept.

"This human self-awareness that has emerged allows two new classes of mental activity that I will talk about in closing. The first is a symbolic transfiguration of our primate minds. The primate, the ape from whom we evolved, had knowledge, nurture, temperament, empathy, hierarchy. One of the things that we do with our symbolic minds is to access our primate minds, but we do it symbolically. An ape has knowledge, but we have the capacity for mindfulness. We experience our knowledge.

"This concept is very important, almost religiously important, for me because it means that I as a human am not disconnected from my evolutionary path. I am a part of it, and yet I am also experiencing my mind quite differently.

"Finally, of course, we have novel kinds of human experience, for which we cannot find antecedents. Chimps are not known to do mathematics. They are not known to paint pictures. We have aesthetic experience. We experience what I call moral beauty, the beauty of moral experience. And we have understandings accessed with contemplative practice.

"I would claim that this capacity of our minds for symbolic reference and symbolic self-reference represents the highest level of emergence, which is a new thing on the planet. It has not been seen before, at least in this cycle."

Compassion: Unique to Humans?

Ursula's presentation had raised many points that led into territory where we hoped to uncover valuable insights from the Buddhist side, and Arthur was eager to open discussion while the material was fresh. "Your Holiness," he began, "now we'd like to pose some questions to you. We would like to ask you your views on the relationship between the animal kingdom and the human kingdom. What's the difference? What are the similarities? Ursula very beautifully pointed out that she honors and finds value in her connection to our animal origins and all the biological kingdoms behind those. But there are also certain distinct human capacities. How do you understand these, yourself, or within Buddhism?"

The Dalai Lama began with characteristic humility, "I haven't given much thought to this in a thorough manner and I am not a researcher, but one could attribute many of the things that we human beings possess to animals. These include thought, a wide range of emotions, and probably some form of memory as well, because they remember to do things that they learn. Where I see the greatest difference is the level of intelligence. The animals can have a degree of intelligence, but the ability to enhance intelligence and use analysis may be what distinguishes human beings.

"Related to this, there is one point I would like to have your opinions on. If you look at the animal world, there seems to be a correlation between empathy and a connectedness. For example, where there is a very intimate dependence between a mother and her offspring, there seems to be nurturing and empathy. But in other cases there isn't. Imagine that a mother turtle who had laid her eggs were to come back and meet the offspring. Probably there wouldn't be any sense of connectedness or any bond. Similarly, if a mother butterfly were to meet her offspring, probably there wouldn't be any empathy or a sense of connectedness.

"Also in the mammal world, the natural bond and empathy between the mother and offspring seems to last only until the offspring becomes self-sufficient. Then the dynamic seems to change. In some cases, one could possibly eat the other. But here the human being seems to be different. Even though the offspring is no longer dependent upon the mother for survival, human beings maintain that bond.

"This raises a question about compassion. Is compassion, or a sense of close feeling, not something moral, but rather a physical necessity? Does it arise because you need it, and if you do not depend physically on another, does that feeling not come? Is there some sort of biological or evolutionary basis for this?"

Ursula responded, "Your first point about intelligence—I would argue that the difference between our intelligence and the intelligence of an ape has to do with our language. In fact, apes are smart about things that we're not very smart about. They can swing in trees. We don't know how to do that. They have a different kind of intelligence than we have. It's not that we took the apes' intelligence and made it bigger, or something like that. Rather, we have a different kind of intelligence that has to do with our ability to use symbolic language. In experiments where people are trying to teach language to young apes, they work and work at it, but there is still very, very minimal ability.

"As for nurture, the turtle and the butterfly do nurture their offspring, but in a very simple way. They build a nest. A sea turtle climbs out of the water, a big thing for her. She is not herself interacting with the offspring, who are left on their own, but there is an impulse to nurture. It's just of a different kind. And even the caterpillar builds a cocoon. Among the primates, long after a baby ape has been weaned and is no longer dependent on the mother for milk, they have a relationship for life. They know who their mother is and who their siblings are, and this is very important in their social group. We were gifted with that from our ancestors."

Arthur stepped in: "Can we address a little further, though, the question of compassion and ask not only about those who have a kinship relationship, but about those who are not related at all? Do animals have empathy or even compassion for unrelated members of the species?"

Ursula answered, "In the chimps certainly, and to a lesser extent in the gorilla and orangutan, there are many examples of empathy. For example, if two young chimps have been fighting, an old female has been observed to come after they are finished fighting. She holds out both hands and takes them, and they calm. She knows how to do this. There are lots of examples: a chimp doesn't know its father, only its mother. But the fathers are very interested in the community and have lots of empathy and behavior that's not selfish, such as sharing food."

"If it's not a matter of survival, where do you think compassion comes from?" His Holiness asked.

"The big difference between animals like us and a turtle, for example, is that we are social. We live in groups, and part of our survival has to do with this. Social animals have evolved ways of organizing behavior so that the group itself survives."

The Dalai Lama tested this argument: "In the case of honeybees in a nest, they rely upon each other for their survival. And they all work, collaborate together, even without having a constitution, or knowing any Dharma, or anything like that. They all work together that way, simply because it's necessary. And their emotions, anger anyway, are for survival."

"Absolutely. Yes," Ursula confirmed.

"And sexual desire is also for survival. And competition is for survival."

"Absolutely."

"And compassion also—"

Ursula completed the thought, "—is for survival. It's not good for the group to have male chimps fighting and so the old female helps them with their anger."

"It seems that this is the one real truth, that feminine souls are more compassionate," the Dalai Lama chuckled.

"But the males will do it too," Ursula continued. "Let me tell you another story. A female bonobo was playing with a bird, and she injured it so the bird could not fly. She looked at it and then she climbed up into a tree, holding the bird with one hand. When she got there, she moved the bird's wings up and down, trying to get it to fly."

"Outside their own group . . ." His Holiness mused. "That is really interesting. But maybe it was an exception."

Compassion and Intelligence

Arthur again stepped in to steer the discussion. "Your Holiness, I want to go one step further. What is the difference between this behavior and that of a human being? Let's assume, for the sake of argument, that the higher primates share with us at least some element of compassion and intelligence. And now here we are, in Dharamsala, having our meeting and discussions, our cultural thing. What is the difference, in your view, between us and these primates that have such behavior? Is it a degree of intelligence? Is it because we improve ourselves through self-education and culture, or through discipline? Or is it just that we innately have more intelligence or more compassion?"

"I think that animals in general, but in particular the social animals, have a limited altruism," His Holiness said. "Again, this is out of necessity. But they can't develop infinite altruism. This only a human being can do, because of intelligence. And also, I think, because of instruction. Animals can carry instructions, but only in a very limited way."

Here Steven Chu ventured a question. "Your Holiness, do you think this is just because we have more instruction, more compassion, more altruism, or is there something that emerges with this 'more'? Is there so much intelligence that a new level appears?"

"It's really a difference in degree and application," the Dalai Lama responded. "It's not a qualitative difference. But even on a physical level, there's a difference in the size of the locus. The human brain is much bigger. But what do we mean by a qualitative difference? Intelligence of a different kind? Of course, if you look at human experience, we have capacities that animals do not have, for example the symbolic language that Ursula has been talking about, or the ability to conceptualize things, to use abstract ideas, and also to relate past experiences, past memories, to present experience. All of these are capacities that human beings have and animals may not have. But to label that as a qualitatively different intelligence is problematic. Intelligence is the capacity to distinguish what is and what is not. As far as that is concerned, primates have this capacity as well as humans."

"In Christianity there is a dividing line," Steven offered. "They say a person has a soul. That's a qualitative difference."

"Buddhism does make that kind of distinction between men and animals," the Dalai Lama said. "Language is the major distinguishing factor, but that, in a sense, is a function of intelligence. I mean, symbolic language. We cannot say that human beings have language and animals don't. Animals may have their own particular languages."

"Oh, they communicate," Ursula agreed. "There's a lot of communication all the way through the animal kingdom."

People of Mind and Life: Richie Davidson, from the University of Wisconsin, board member of Mind and Life.

"One of the previous Karmapas could give teachings to birds," the Dalai Lama said, with a nod to the present Karmapa sitting among us. "I have no such capacity. In birds, some calls are the sign of danger, a kind of warning. Some calls are a sign of relaxation, and some are the expression of desperation."

Ursula brought the discussion to a close with a tantalizing vision of future research. "As you know, Your Holiness, there are laboratories, like that of Richard Davidson in Wisconsin, where scientists are looking at the minds of monks in meditation.

"These same kinds of technologies will be soon used to look at the difference between a chimp's mind and a human's mind. Right now we can't do this because the chimp does not want to be in a machine. But when the machines become less cumbersome, so that a chimp could just wear one and still walk around, it will be possible to ask whether their minds are working the same way as ours. We will see then whether ours is 'more' or whether it's just different. That will be very interesting."

Reflections

Ursula had mapped a very long path reaching from cell structure to human culture. It is worth emphasizing the homogenous and graceful flow of biological evolution, which is indeed the force and beauty of the Darwinian interpretation of life. Several important concepts have appeared in this descriptive arc. The origin of the flagellum epitomizes one principle: nature itself does not want or need its new developments, they just happen. There are no big plans in evolution. There are modifications that arise according to the dynamics of contingency, and if one modification is useful for the reproduction of life, this becomes "necessity"—using the old terminology of Monod. In other words, it becomes part of the invariance of the genetic code. The importance of contingency has been emphasized also in the previous chapter on the origin of life, and it appears to be an important contributor to the history of our planetary life.

Is there then no "finality" in a flagellum, or in the human brain? The question of finality or teleonomy in nature is very complex. However, the underlying observation is that the notion of finality needs an observer, the human observer, to impart value judgments to the things of nature. There is no finality per se in an amoeba swimming in a sugar gradient: the finality is attributed by the observer. The amoeba and the bacterium do not know that they move according to the laws of chemistry, physics, and biology.

Discussions continued during lunch. Matthieu Ricard and his mother converse with Richard Gere.

Are our observations on the origin of the flagellum also true for the origin of the human brain? Is it a product of contingency, a hard picture only moderately alleviated by the notions of self-organization and emergence? And is it also true for the products of the human brain—human culture?

Most scientists would answer this question positively—keeping in mind, however, Francisco Varela's "embodied mind," the notion of a melting congruity between mind and body, in the sense that there is no human life without consciousness and there is no consciousness without an organic living structure. The two are one in the process that science calls life. Buddhism, however, has serious reservations here, which the Dalai Lama would elucidate in due course.

An Interview with Richard Gere

Interviews were usually held at the Chonor House, where we had our lunch. This was in the form of a buffet, and after moving through the line to fill your plate, you would sit at one of the small iron tables in the garden. These tables were also where animated conversations and debates over the morning session would begin and continue. We were hungry and we did not talk for a while. Then I began:

LUIGI: Do you recall the moment when you started on this path of Dharma?

RICHARD: It was fairly early for me. I always had an interest in philosophy, and I had quite a religious upbringing as a Methodist, as my father was quite involved with the church. However, I found a great limitation in the Methodist Church, and Christianity in general, when engaging with various fundamental, serious questions. When I went to college I was a philosophy major, and I was struck by one philosopher in particular: Bishop Berkeley, whose thesis was "subjective idealism," which held fundamentally that reality is a function of mind. I realized later that it was essentially the Mind-Only School.

Like most young men, I was searching as a teenager and in my early twenties. I was not particularly happy and I wasn't quite sure why. I started reading a lot of books. I withdrew into myself and read a lot of traditions. I gravitated to the Eastern traditions. The first one that really struck me was Zen Buddhism. By the age of twenty-two or twenty-three, I was just beginning to practice *zazen* in New York. I had studied with a few Japanese masters and was totally fascinated with that world, especially the Japanese version of it.

LUIGI: Was mostly your brain curious, or did you find an internal correspondence?

RICHARD: There were three aspects. Intellectually, it was stimulating to me in its willingness to engage the serious questions. Emotionally, clearly, I remember the first times that I meditated were incredibly emotional for me. It still is to some extent, but certainly in the beginning there was a well of emotion. I had been aware of its taste, but I really had not seen it, and I saw the face of my emotions at that point. Aesthetically, it also appealed to me, especially the Japanese aesthetic. The color, the form, the rituals had enormous appeal for me. They spoke to me not in a dry way, but in a very emotional way, the way art speaks to one aesthetically. It touches your heart.

When I was about thirty, I had a very strong yearning to go to Tibet. I had read some Tibetan material, not specifically by His Holiness but just general Tibetan material: mostly the Evans-Wentz books—the earlier translations were very powerful—and the Milarepa material, but other things too. At that point a friend of mine, John Avedon, had written a book called *In Exile from the Land of Snows*, which was a political history of Tibet before and after the Chinese invasion. And he said, "Look, if you really want to meet Tibetans, don't go to Tibet. Go to Dharamsala." Of course I had this romantic idea that I was going to find my teacher in Tibet, but I said fine. So I came here to Dharmasala. and left a card-carrying Tibetan Buddhist.

LUIGI: I had a similar story coming from Zen. Did you have the same experience as I did of the difference between Zen, this rigorous, almost Protestant view of Buddhism, and the rich, baroque, almost Catholic flavor of Tibetan Buddhism, with its *tangkas* and its many gods?

RICHARD: Yes, exactly, and it takes a while to penetrate that and see beyond the surface of it. You need good teachers who can fill out the symbolism of what you're looking at in terms of color and shape and deities.

LUIGI: The aesthetic of Zen is very different from this.

RICHARD: It's a very strict, very Calvinistic approach, which I still like enormously. But I've also come to appreciate Tibetan Buddhism for what it is, and find the joy and richness in it. In many ways this is missing in the Zen approach. I don't think I've ever laughed as much in my Zen career as I have in my Tibetan career.

LUIGI: You described yourself as a young man who was not very happy, a searching one.

RICHARD: I was extremely unhappy, and certainly I had considered suicide.

LUIGI: And now you think of yourself as a different person in that sense?

RICHARD: Oh, absolutely! I could see what led me to that point of suffering. I could see logically as well as emotionally why I was in that state, and Buddhism very clearly gave me the tools to get out of it. My early reading of Buddhism essentially came out of the existential world—I remember carrying around *Being and Nothingness* when I was a teenager, of course not knowing what it was but having some instinctive connection to it. At that time the notion of emptiness to me meant nothingness. If I could achieve this state, I would disappear, and therefore, there would be no one to suffer. I would just disappear and the universe would disappear with me. Of course, my ideas have matured a bit since then, including my understanding of what emptiness really is. Not that I have any realizations, but I certainly have somewhat of an intellectual understanding.

LUIGI: Is your Buddhist internal life in conflict with your Hollywood life, or do you find that they are in harmony?

RICHARD: Clearly, that's a job that I do and that I like very much. But we all have a job. The essential thing is that we're all the same. We all have the same problems of ignorance, hatred, desire, and jealousy—the whole arsenal of afflictive emotions. And that's the issue. The superficial things of where you live, what your job is, whether or not you have money, whether or not someone wants to take your photograph are totally irrelevant.

LUIGI: Are you saying that your job is not as important as the search for wisdom?

RICHARD: They're not mutually exclusive. Certainly everything that I do in my job tests every teaching I've ever had as a Buddhist.

LUIGI: Do you mean that you make a test in whatever you do?

RICHARD: The study of Buddhism, the practice of Buddhism, is essentially watching your mind. And when I say "mind," I mean mind/heart. To interact with the world, to have a job, is to have emotions constantly arising in the mind. It's an extraordinary opportunity to see what really is the color of your mind, the color of your emotions, where you're really coming from, what you really have achieved or not achieved. Constantly you are in a situation of having to manifest love and patience in areas where maybe before there was no patience, or there was anger. That's bouncing off of life. That's why for us modern people it's an extraordinary opportunity to practice in a really vital way. The mirror of life is very, very sharp and clear for us. The cave is not our life.

LUIGI: Can we say that mindfulness is the substrate or the background on which you operate as a successful artist?

RICHARD: The word "artist" to me is a loaded word. I think we're all artists. The art of life is the one we're all students of. But as an actor, it's a very interesting profession in terms of being a Buddhist practitioner also. Again, you're looking at the mind. You're playing with emotions. My job is to use my emotions to illustrate the human dilemma. Clearly, I have to know my own emotions to be able to use them in an effective way in my job. So I'm doing really good practice all the time if I can see it that way. I'm really doing good operatic practice constantly in my job.

5 / The Magic of the Human Genome and Its Ethical Problems

AND AN INTERVIEW WITH HIS HOLINESS THE KARMAPA

IT IS NOW TIME to move one step further up the ladder of complexity of life and consider genetic heredity. One of the most basic laws of nature is the conservation of species: from one generation of roses arises the next generation of roses; elephants produce elephants again and again; and so it is for fruit flies and green peas and daisies. Heredity became a scientific question in the nineteenth century with the acceptance of the view, by around 1860, that all plants and animals consisted of cells, and that cells give rise to new cells through division: *omnia cellula ex cellula*. At that time, only intuition related this cell-to-cell transmission with the mechanism by which a sperm fertilizes an egg.

Was there something in the cells that gave rise to the constancy of the species? Or was a mystical force linked to the sacredness of life itself?

The first important clue toward a scientific answer came with Frederick Miescher's discovery in 1869 of the material in the nucleus of the cell that would later be recognized as DNA. By the 1880s, some scientists began to suspect that this material was linked with heredity. Advances in light microscopy enabled biochemists to observe that the sperm penetrates the egg and that the nuclei of the sperm and egg fuse. It was Walther Fleming who observed the cell dividing and chromosomes replicating, and concluded that chromosomes were a source of continuity from one generation to the next. Meanwhile, the Austrian monk Gregor Mendel was tending his pea plants, making observations that remained in obscurity for years before eventually being accepted as revealing the logic of genetic heredity. But we will leave that piece of the story to Eric Lander's telling.

Arthur Zajonc introduced Eric's presentation on this morning devoted to the human genome: "One of the most dramatic new developments in the study of genetics is our understanding not only of the molecular mechanisms themselves but of the detailed set of instructions that the DNA molecule offers for the development of particular biological traits. Not only do we understand the simple genome for elementary organisms, but at the laboratory of Eric Lander and his colleagues, for example, extraordinary work has been done to decipher even the human genome. What this offers is a possibility not only of affecting the niche in which we evolve both as biological and as cultural creatures but also of directly affecting, in a single step, the biological basis for life. It's about this aspect that Professor Eric Lander will be speaking to us."

The Text of Life

Eric began by explaining his motivation in joining this dialogue—a respect for the deep ethical principles of Buddhism, and the hope that they might offer some guidance in the ethical dilemmas that now confront us in the West as a result of advances in genetic science. But first he would explain the science underlying the choices we now face.

"We understood from Ursula's presentation that in each individual human there are billions of cells, and each of these cells has within it billions of molecules. These molecules include the DNA, a sequence of four building blocks that have their own chemical shape. We just call them A, T, C, and G, because it is easier than writing the chemical shape. And so we may think about the genome as a text. Every single cell in your body has the same text.

"To explain this genome text, let me compare it to a text that may be more familiar. As Thupten Jinpa has taught me, the Buddhist scriptures and commentary—the Kangyur and the Tangyur—are at least about 300,000 pages altogether. I am guessing that is about 450 million letters. Now, of course, you can print this in a modern way, and even 600 years ago it was printed from wood blocks. But before that, it must have been copied by hand. Usually, I'm sure, people would be very careful, but sometimes mistakes were made.

"Do copying errors matter? What about one letter? Can an error in one letter matter very much in a text? The answer is yes. As I understand it, at the end of a sentence, you may have 'yin,' which is an affirmation, or 'min,' which is a negation. It only takes one letter to change 'yin' to 'min,' but it can change the entire meaning of a passage. Of course most letters may not have such a big effect, but some letters will. The same is true of DNA.

"When a monk makes a mistake in copying, maybe the next monk who copies that book will copy that mistake and add some of his own mistakes too. If we had hundreds of copies of the Kangyur and the Tangyur, we could even tell that one copy must have been copied from another because it is very similar but has a few changes, or that another one must be rather distant because many more changes have been introduced. We could reconstruct the history of the copying.

"This is very similar to the way we compare DNA texts. How big is the human text? It is about two million pages, or three billion letters. That is how much information every cell in our body has. We have just come to have this text in the last year or two, and you may ask how well we have studied it. Well, you have had your text for a lot longer, and I imagine you have the sense that there is always much more to be learned. We certainly have a sense of humility about how much we know. Nonetheless, I will tell you about some of the things we do know, but I am a beginning reader.

"In addition to having these two million pages, we actually have two copies of this text. We get one from our mother and one from our father. And when we have children, we pass on one copy, which is actually a mixture. We take a page from the father, and then a page from the mother, and another page from the father, and so on. And that is what we pass on to our children. Do we pass it on perfectly? No. Out of three billion letters, we make about fifty mistakes in every generation. It's pretty accurate, but still sometimes there's a mistake like 'yin' to 'min.'"

In planning his talk, Eric had made a real effort to find analogies that would be familiar to the Tibetans, but the Dalai Lama was one step ahead of him here. "When you talk about a mistake in copying, do you mean a mutation?" he asked.

"Exactly. By mistake, I mean what a geneticist calls a mutation."

Eric continued, "So we have this text, this genome, in every cell, but it is used in a very physical sense. One passage of this text may be used as the instructions to make a protein that is in my hair. Another passage may be used to make the protein in my blood called hemoglobin, which carries oxygen around my body. We think there are at least 30,000 different passages that contain the instructions for making these building blocks, these proteins, that are used in many ways. Together they make up the different cells of our body.

"But of course, they do so together with the environment. Whether or not a person will be very tall depends on nutrition as well as genes. Whether or not a gene will be turned on also depends on environment. For example, at lunch today, a particular gene will be turned on in my cells that helps digest

Eric Lander leaning forward and explaining his "book." Arthur Zajonc
on the right.

sugars. In fact, it's on a little bit now, but after lunch it will make ten times as
much product in response. If I exercise very hard, it will affect the expression of genes to produce more proteins in my muscle. Mental activity will
also affect the expression of genes. Depriving an animal of mental activity,
or providing a very rich stimulating environment, affects the growth of brain
cells, and that is dependent on the turning on and off of genes.

"So that is my notion of the genome as a text. It is an imperfect analogy,
but it is a helpful one, for me at least. And I hope, given your tradition of
texts, it may be a helpful one for you too."

Mendel's Peas

At this point Eric Lander set off on a rich description of the early history of
genetics, starting with Gregor Mendel in the early nineteenth century. Eric
reminded us that it was not only simple curiosity that started the Austrian
monk's investigation. People by then had realized that the possibility of
breeding plants and animals brought back to Europe from other parts of
the world could have tremendous economic importance, and easier trans-
portation to wider markets also provided an economic stimulus for research. Organizations were formed to promote the study of heredity, and
one was led by a man named C. F. Knapp. Knapp also happened to be the
abbot of a monastery, traditionally a center of learning. "He began to look

for young students who had training in physics and in mathematics to enter the monastery," Eric explained, "so that in addition to their monastic duties, they could also take up scientific problems for the betterment of the world. His best student was Gregor Mendel, and the hope that he could make a difference for the betterment of the world was, I think, realized in many ways."

In what is considered the beginning of modern genetic science, Mendel bred pea plants and tracked the heredity of various traits—the height of the plants, the color of the peas—over generations. He realized that when he crossbred plants with different traits—tall with short, or green with yellow—all of the first generation of progeny shared the same trait. They were all tall plants, for instance, or all produced green peas. But when he crossbred plants of this generation among themselves, the missing traits reappeared, and in a consistent ratio. For every three tall plants, there was one short one; for every three that produced green peas, there was one that produced yellow.

"And so Mendel made a beautiful application of pure logic," said Eric. "He thought: *What can be going on here? There must be a hidden element that continues yellowness.* Mendel saw that these results could be explained if the traits were inherited as elements—what we now call genes—of which each plant received two from its parents and passed on one to its progeny. Thus traits that were apparently missing—the short plant or the yellow peas—might only be hidden, or what we call recessive."

The importance of Mendel's observation did not became clear until the end of his century, when both he and Darwin were dead. Darwin would have benefited enormously from Mendel's discovery, but in fact, Darwin himself had no idea of the physical basis of heredity.

"But as so often happens in science," Eric continued, "three different groups around the world rediscovered Mendel's ideas at the same time. The twentieth century opened with the publication in January 1900 of the first of the papers describing the rediscovery, and to my mind, the twentieth century is the century of genetic science.

"The next twenty-five years were devoted to finding out where in the cellular substance heredity resides. By the end of the first quarter of the century, we knew that it is contained in the nucleus at the center of the cell, in the structures called chromosomes. The term 'chromosome' means 'colored thing' and is a statement of how little we knew at the time: only that a colored dye would stick to it."

* * *

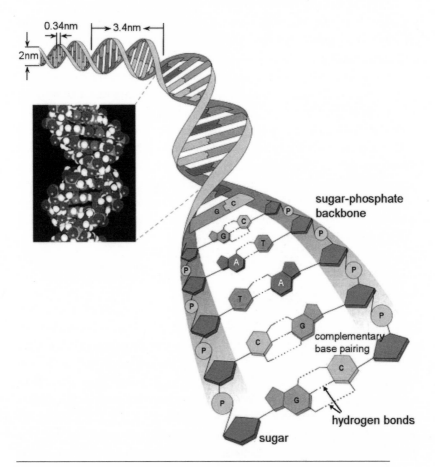

A schematic diagram of DNA's double-helical structure. Note the four bases, indicated as A (adenine), G (guanine), C (cytosine), and T (thimine), complementarily bound to each other (A to T and C to G) by hydrogen bonds. In the insert, the three-dimensional rendering of the double helix.

MORGAN AND HIS *DROSOPHILA*

From the beginning of the twentieth century when genetics became a science, the basis of this new discipline was the collection of data about the heredity of traits and their link with the chromosomes. At this time, the famous fruit fly, *Drosophila*, came to the aid of science. With a life cycle of fourteen days, it would breed and therefore permit more observations in a shorter time than plants or higher organisms. *Drosophila* has the additional advantage of having only four chromosome pairs, simplifying the statistics. It was the American geneticist Thomas Hunt Morgan who first explored the genetics of *Drosophila*. His basic observation was that genes are strung together in a defined linear order in the chromosomes, and that some genes remain linked together while other cross more frequently.

Another important concept was introduced by a student of Morgan, Hans J. Müller. He recognized that genes have the capability to mutate, changing their characteristics. Using X-ray irradiation, Muller was able to increase the mutation rate of *Drosophila* by 15,000 times over the natural rate, and produced some monstrous flies with misplaced wings or enormous eyes—the first experiments to artificially create novel beings. He realized that the mutations were due to chemical changes induced by irradiation, but no one yet understood that such mutations would also form the basis for Darwinian selection.

Why the Human Genome Project?

"The next twenty-five years, until the middle of this century, were devoted to understanding the molecular structure of chromosomes," Eric continued. "This is when the discovery of the structure of DNA, the famous double helix, was made by Crick and Watson. The next twenty-five years after that were spent learning how the sequence of the nucleotides actually specifies the proteins, how they are read out three at a time, how they are translated into the sequence of the building blocks of the protein.

"And, in addition, these years were devoted to developing technology to be able to read these nucleotide sequences ourselves. By about 1978, we could read a few hundred letters at a time. It's not much, maybe a sentence or two. And we could take the DNA from a cell, chop it up using chemical tools, and place it into bacteria. The bacteria would then reproduce and make copies for us of this or that sequence. We might have different bacteria, each carrying one phrase of this text. Now we can take that bacteria, purify out the DNA, and perform a chemical process on it to read the instruction.

"The motivation for this work was medical—to try to understand the inheritance of disease," Eric said. The idea that hereditary diseases such as cystic fibrosis, sickle-cell anemia, or Huntington's disease are due to a "wrong" DNA sequence has long been established, but finding the particular sequence at the root of a disease is a long and arduous process. Eric talked us through it, using the example of a family where one parent has a disease that is inherited by half of the children—comparing the DNA of each family member, looking for a particular mutation that should appear only in those individuals with the disease. In theory, the logic is beautifully simple, but in practice it is like trying to find, in Eric's analogy, a single letter in a huge book whose pages are unbound and scattered on the floor.

"If I find the part of the book where the inheritance of a particular letter matches the inheritance of the disease, then I know that the cause of the disease must be near that part of the text. How do we find the gene itself? It may be two hundred pages away, which is still very close in our very big book. What we do is start with the letter that corresponds to the pattern of inheritance, and we take that page, and we go walking around looking for where the next page is. Maybe there's a little bit of overlap, so we can find it. Then we have to walk around to find the next page, and the next.

"It is really boring, and expensive. It took five years and $50 million, but at the end, the right page could be found. In this way, the gene was found for cystic fibrosis, a disease that causes thick mucus to build up in the lungs. There's more, though. We can take this sequence and translate it into the corresponding protein. Then we have the computer search for all other proteins that have ever been studied until now, looking for a similar sequence. And the computer finds that it is like lots of other proteins that sit on the surface of the cell and transport things in and out. In an instant, you discover that you have probably found the function of this particular gene.

"This is very powerful because we have connected knowledge across all of science. And of course, this knowledge can be put to use in developing treatments and diagnostic tests, screening for risk of inherited diseases. It sounds good, so what's the problem? The problem is that it takes five years and $50 million for just one disease. This is crazy. And so the Human Genome Project began as the idea to do it once and get all of the book sewn together. In this way a researcher who wants to study a disease could just look up DNA sequences in the book.

"It took about fifteen years to collect our text. It involved twenty different laboratories in six different countries: the United States, the United Kingdom, France, Germany, Japan, and one lab in China. The largest of those labs are in England, at Washington University in St. Louis, and at my own center, the Whitehead Center for Genome Research at MIT."

In fact, the project was spurred on by competition between academic research and private enterprise in the form of the Celera Genomic Corporation led by Craig Venter. Venter had invented several techniques that speeded up the process of sequencing DNA, and he launched the project to decipher the human genome with the aim of patenting it and making money. He offered a subscription database and shared data for free only

Thupten Jinpa clarifying the notion of the human genome to His Holiness.

with scientists who were willing to commit not to propagate it. But the notion of patenting the human genome struck a nerve in academia, which led to more funding and the vast collaborative effort that in the end prevailed, and which Eric was describing to us now.

"All of these centers worked together in a consortium over the course of many years with one commitment, which was that all of the information must be made available every day with no restriction. Basically, it had to be put up on the Internet before we went home at night. In February 2001, two scientific papers came out reporting about most of the sequence. It was not completely finished, but as of today, we have about 97 percent."

More recently, a new three-year, $100 million endeavor, called the International HapMap Project, has been launched. To understand why some of us are more likely than others to succumb to particular diseases, geneticists need to sample populations that are affected by one disease and study the individual mutations known as single nucleotide polymorphism (SNP). If a particular SNP is inherited with the disease, this is taken as a strong indication that a gene that confers that disease lies somewhere nearby. The HapMap, based on the analysis of blocks of sequences that have been inherited for generations without being broken up—the haplotypes—should provide geneticists with the means to scan the entire genome rapidly for disease genes. We will come back to these questions later, but now it is time to let Eric continue.

"Some parts of the text have lots and lots of genes—meaningful sentences that make proteins. And of course we want to study each of those

genes and understand what they do. Some parts of the text have many letters but relatively few meaningful sentences, as far as we can see. We don't
fully know what these letters are doing, but we have good reasons to think
that they have no meaning, or perhaps a very different kind of meaning."

Although it took a long time to decipher the entire human genome,
more rapid progress has now been made on other animals and plants, as
Eric explained. Genomes for the mouse and three kinds of fish are complete, with the chicken, dog, cow, and chimpanzee in progress or soon to
come. "And so the animals and plants are standing in line now, and we
know what kinds of books we will acquire for our library, at least for a few
years."

Eric still had much to say about the comparison of different species' genomes, but it was time for a tea break, so he closed this section of his presentation on a very gracious note. The human genome, he said, "is very
much the heritage of all humanity, and it was therefore important that
many different countries worked on it." At the research centers in six different countries, there were people from many more countries involved in
the collaborative effort. In fact, Eric said, there were forty Tibetans working with him at the Whitehead Institute at MIT. Most of them had lived in
Dharamsala at one time, and were very excited when they learned of Eric's
invitation to the Mind and Life conference. And so now Eric relayed their
greetings, letters, and a white *katha* scarf to His Holiness. The Dalai Lama
smiled and nodded, signaling a break. A stretch, a cup of tea, a few moments of conversation, and we were ready again to listen to the next episode of Eric's story.

Comparing Texts: Of Man and Mouse

"We now have the DNA text for the human, and we have the DNA text
for the mouse. How similar are they? It's a complicated question. What
do we mean by 'similar'? In terms of the size, they're both very big." Eric
returned to his previous analogy: if the human genome is six times the
size of the Buddhist scriptures, the mouse genome is five times that size.
"But if we look at the text very closely, we can compare the text and see if
the same strings of letters and even the same sentences occur.

"What we mean by 'similar' is that there are many parts of the text that
are extremely close. Perhaps 100 letters here and 80 letters there are the
same. That is much more than by chance. And we see that again and again,
and they occur in the same order. What we infer is that there was a common text about 100 million years ago. This text has been copied, and it

changed a little bit, and it changed more and more with time. What we see is what is left after that process of change, but it is still clear that these must have originally been the same text.

"Now if we look across the whole human text, we see that it does not go in one continuous string that we can line up with the mouse. But we can take the text and say that this part of the mouse text corresponds to a particular part of the human text on chromosome number 2. This other part of the mouse text corresponds to a particular part of the human text on human chromosome number 8. In addition to changes in letters, there has been some rearranging of parts of the text. One of these breaks and rejoinings happens about every half a million years, on average. But only about 300 or 400 breaks need to be made to turn the mouse order into the human order, or the human order into the mouse order.

"So this is how we infer some of the history. It is an inference, but we think it is a strong inference because it seems unlikely that these texts would be so similar unless this process happened. And when we compare these texts between different animals, [our inferences] tend to agree with the inferences that we draw from fossils.

"What about a much more distant comparison, between a human being and yeast? Surely, that's pretty far. We think that the common ancestor between the human being and the yeast lived about 1.5 billion years ago, and yet, when you look at the text, there are sentences that are almost identical, although not that many. Some of those sentences are the genes that have the instructions for making the cell divide. When mutations occur in those genes, they can prevent the cell from dividing or cause it to divide too much—they will cause cancer. We can study cancer in the yeast be-

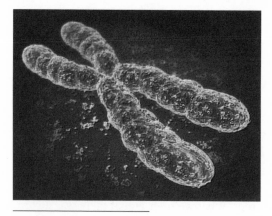

The human chromosome 8.

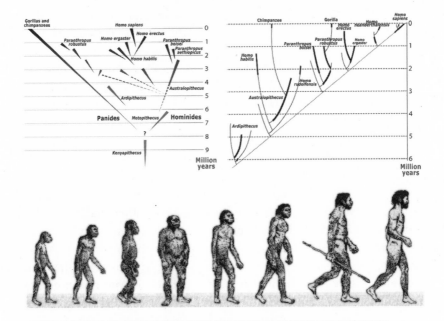

The evolution of *Homo sapiens*. In the higher panels, two different theories (among many) of human evolution; and below—rather in scale—the corresponding evolution of hominids.

cause the basic machine that causes the cell to divide has the same components as in human cells. If we take the yeast gene and put it into a human cell—not a human being but a human cell—in many cases the human cell can use the yeast gene. So we know that the sentence has largely preserved its meaning. This is true between a human and a yeast, and a human and a tree. We think these parts of the text are very old. Of course there are many other newer parts of the text, like the hox genes involved in developing an embryo, that you won't find in yeast. And that makes good sense."

A Small, Antique Family

"A comparison between the genomes of two people shows that they are much closer, of course, than the human and the yeast or the human and the mouse. How much variation will there be between any two people in the world—for example, a Tibetan and a Swede? The difference is about one letter in a thousand. This is true also between two Tibetans, or between a Tibetan and an African. The level of difference is essentially the same between any two people on this planet. It is a very small difference.

In fact, the difference between a typical Tibetan and a typical African is smaller on average than the amount of variation among Tibetans. There is more variation within an ethnic group than between different groups. If we compare two chimpanzees, how close are they? The answer is three letters in a thousand. Two human beings anywhere on this planet are more similar to each other than two chimpanzees are to each other."

The Dalai Lama jumped in. "That's all the more reason that human beings should be more harmonious."

"I agree. This is exactly my point: compared to most species, in a way we are a very small species. What does it mean that we have so little variation? It means that we used to be a very small population. A population that is large will have a lot of variation, while a population that is small has very little variation. We have six billion people, but this is very recent. In fact, three thousand generations ago, all human beings trace back to a founding population in Africa that we estimate had about ten thousand individuals. In a population of ten thousand people, you expect genetic variation of about one letter in a thousand. We show today the same level of genetic variation that we had in Africa. Three thousand generations are not enough to acquire a lot more, because new variations come very slowly. Most of the variations that we find in Tibetans are the same that we find in a village in Africa. This is true all over the world, and it is one of the reasons we should not think about the differences between ethnic groups. In fact, we are very closely related as a family. We can use this information to trace our origin back to Africa and also, interestingly, to trace how people left Africa and migrated around the world."

The emphasis on the absence of significant genetic differences between various human populations was an important point for Eric, and he knew that the scientific demonstration of the nonsense of one ethnic group claiming genetic supremacy over others was very important for His Holiness as well. The ethical problems that arise from the sampling of DNA from different populations and ethnic groups bring to mind the troubled history of the Human Genome Diversity Project (part of the Human Genome Project). When this project was proposed in the 1990s, its founders were originally interested in anthropological questions. They wanted to analyze the DNA of different ethnic groups and link the findings to studies of language and other cultural features. As Carina Dennis describes,[1] many groups representing indigenous peoples protested, with the argument that this would exploit vulnerable populations and interfere with people's beliefs about their origins. The project was dropped.

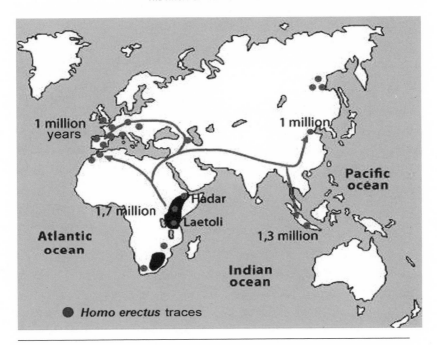

Migration chart of *Homo erectus*: starting from Africa and moving to Europe and to Asia.

Genes and Diseases

"Genetic variation can 'cause' disease," Eric continued. "I put 'cause' in quotation marks here, because geneticists, like Buddhists, are mindful of the fact that there are not single causes. But sometimes there are very strong causes, and I will give you an example of a strong, but not complete cause." Here Eric offered some complex charts of chromosomes to show that certain people have a characteristic "spelling" in their genes that allows us to predict that they have a higher probability of getting Alzheimer's disease.

"This poses many interesting questions. Do I want to know what 'spelling' I have? Personally, I don't want to know, because right now there's nothing I can do to prevent Alzheimer's. If I found out that I was in this group, I would be emotionally in turmoil, and I don't see the benefit of that. But I have friends who work in academia and in the pharmaceutical industry who have figured out what protein corresponds to that gene and are trying to make medications that will prevent, or at least slow down, the process of Alzheimer's disease. When they succeed, I will want to know."

"That's very wise," said His Holiness, and Eric beamed at the compliment.

"We will have many choices like this. Let me tell you about some more examples of variations in our population. Take the ability to digest milk: some people cannot drink milk because they cannot digest the particular sugar called lactose. This is because, as adults, they do not have the enzyme that digests that sugar. As babies they had it, but the genes turned off, and they no longer have that protein.

"When agriculture began, people began to keep animals that produced milk, and this became quite advantageous. So we find that in some populations of the world, most people have a mutation that allows them to digest milk as adults. Last year, some friends of mine found this gene, and now they are studying different populations and looking at the history of this change."

For his next example, Eric turned to the drinking of alcohol, not something very familiar to our audience of monks. "Some people digest alcohol very slowly and others digest it very quickly. There are two forms of a gene, and one is less active in some people than in other people. We don't know why that should be, but we know the differences are present in all populations but at different frequencies. In Asia the form that is slower at digesting alcohol is more common."

"Is it true that Native Americans also have a very low alcohol tolerance?" asked His Holiness.

"Yes, and in fact Native Americans migrated from Asia. Many of the gene spellings that are common in the Native Americans are similar to those in Asia." Eric paused, and then offered another example: "Perception. Some people are color-blind. One man out of every twelve cannot perceive the difference between red and green."

The Dalai Lama, who has often raised questions about color perception at our meetings, seemed surprised at how common this was. As we digressed into an informal poll to see how many in the room shared the condition, Eric took the opportunity to point out that there are ethical concerns around even asking such simple questions, and the information should be offered voluntarily. But what the Dalai Lama really wanted to know was what Eric described as "a philosophical problem. What do they see?" He could explain the genetics and the resulting lack of receptors in the eye for red or green, but describing the experience was harder. "I have a friend who has this condition, and I try to ask him what color looks like to him. But we cannot have a conversation about it because we have no terms for discussion." All we can say is that perception will differ according to these genetic variations.

Eric was ready to move on to his next example, AIDS, a problem throughout the world. "There are many efforts to protect people against AIDS, and one of the issues is to understand scientifically how the virus attacks us. One of the very interesting clues is that one person in a hundred is resistant to AIDS. And we know why: there is a specific gene mutation. This gene is broken, in the sense that it does not make its protein. The lack of this protein, as far as we can tell, does no harm, and does much good, because the protein is needed by the AIDS virus as a way to get into the cell. Scientists are now working to see if they can make a drug that, on this basis, will help prevent AIDS.

"We could ask the same thing about any disease, for example leprosy. We don't know about a gene here, but we do have some medical observations. In the south of India, most people have been exposed to the pathogen that causes leprosy, but they don't develop leprosy. Only about one in two hundred people who get exposed becomes infected. We don't know why yet, but I think it may be a similar story. If we could find the mechanism, we might be able to use that information to treat or to prevent leprosy.

"We do know something about heart disease. This is an interesting story: one person in a million will get heart disease when they are a teenager. This is due to fat that builds up in the blood vessels that go to their heart. It's something you might expect to see in a seventy-year-old person, but here you see it at the age of fifteen. It is a rare genetic disease that causes this. Two colleagues of mine, Joe Goldstein and Mike Brown, studied these very rare individuals and found out why. There is a certain receptor on normal liver cells that takes up the cholesterol. Because these people lack this receptor, there is too much cholesterol in their blood.

"That posed the question, can we help other people with that knowledge? For everyone else who has the gene and the receptor, maybe we could find a way to make the gene turn on more and take out more cholesterol. And in fact, drugs were developed that stimulate the body in such a way that the gene makes more of the protein and takes up more of the cholesterol. In the United States, several million people take these drugs, and they have had an effect on reducing heart attacks. Something that was extremely rare taught us a lesson that is now being used by tens of millions of people.

"Cancer has a similar story. Some women get breast cancer very early in life, and it turns out that this too clusters in families. Two genes have been found, each of which, when they mutate, makes a person much more likely to have breast cancer. We don't have a medication that will prevent it, but such women may go for X-ray screening at a much younger age, and more often, so that the cancer will be found much earlier." Eric then touched on several other forms of cancer, about which less is known.

"How is that you have more knowledge of certain types of cancers than others?" His Holiness asked.

"That's a very good question. There are several reasons. For one, some of the problems are inherently easier, and some are inherently harder. Sometimes there may be only one gene involved. Other times, there may be three or four genes together, and this is harder to find. There's also a second reason. Some of these cancers affect many people, and so get more attention from doctors. Some affect few people and get less attention. And there is a third reason. For some of these cancers, pharmaceutical companies can make more money, so there are economic reasons for people to invest in the research. But we have only had the ability to study this for maybe ten years, and so we have only begun. In most cases we still have more ignorance than knowledge, but in the next ten or twenty years, I think we will fill in this picture a tremendous amount more. That's why we must address some of the ethical questions now, because we will be much further along with this program in ten or twenty years."

Ethical Problems in Genetics

"Let me give you some strange examples." Now Eric Lander embarked on a description of some very interesting cases of genetic phenomena that raise ethical problems. "There's a famous athlete from Finland who won an Olympic gold medal for winter endurance sports, a competition that required many long, complicated events in the snow. He had tremendous stamina, and interestingly, so did some other people in his family. A geneticist in Finland studied this case and found that the athlete made a lot more red blood cells than a typical person, and so did the others in his family. When the case was studied harder, another receptor was found. There is a receptor protein in normal people that gets a signal to make red blood cells, which is sometimes on, sometimes off. For this athlete, that receptor protein was always on, signaling to himself to make more red blood cells." Eric explained that athletes can now take a chemical substance that stimulates that receptor. To do so is against the Olympic regulations, which do not, however, exclude an individual with this genetic anomaly.

Eric's next example was a genetic predilection to violence, a subject that is fraught with ethical problems on many levels because there is a tendency to blame violence on genes in a way that has not been substantiated by science. "You know that when you get scared, you feel a rush, and that is adrenalin being released. You also have an enzyme that breaks down adrenalin. There's a rare family in the Netherlands where that enzyme is missing, and they cannot break down the adrenalin. Those individuals, in fact, are mostly in jail for burning down buildings, for rape, and for murder, because they are unable to control themselves. In this case, fortunately very rare, there is a physical cause that makes it impossible or very difficult to control afflictive emotion. This is not to say that those people have no control at all, but they start from a very different position than the rest of us.

"Depression is another example, not rare but common. We don't yet know the genes involved, but we have strong reasons to believe that there are genes that make some people more susceptible to depression than others. And if we can make a medication, then we must ask who should take it. I find this question interesting because here we cross the boundary between normal disease and the causes of afflictive emotions. Genetics will have something to say about these causes, although not quite yet. Check back in five years, and I'll tell you where we are.

"What is the motivation for doing all this? Well, the good motivation is to understand diseases and prevent them. Geneticists now feel that maybe we should try to study every single common variation in the human species, to see what the role of each variation is in disease. There are only about ten million different human genomes, but ten million is not as big a number as it used to be. Ten years ago, we'd say it was impossible to study all ten million variations. Now we just say it is very hard and expensive. Ten years from now, it may be very feasible."

Here Michel Bitbol asked for clarification of an important point: "You said before that the difference between individuals consists of only two letters in a thousand. So how is it that there are ten million different human genomes?"

Eric explained, "Although you and I might differ by one letter in a thousand, if we include two more people, then there might be two or three letters in a thousand where we all differ. If we include everybody in this room, there will be a larger number, even if we exclude the rarer variations. If we consider all the people of the world, we come up with about ten million common variations. The details of the calculation are not easy, and in fact, we have only known this number for the past two years. Many people in genetics, including myself, are working on a program to look at all variations and examine how they relate to all diseases. This is still in the future, but not so far in the future.

"I'm going to talk now about the genetic technologies that allow us to experiment and change, because there are ethical questions here too. We have the ability to remove a gene. It's hard work, but not that hard. You take cells from an animal such as a mouse, grow them in a dish, and add to them a certain DNA text that will possibly cause change in some cells. Then you must identify one of the cells that has changed, and grow that cell into a mouse. This can all be done, and in fact, a graduate student will routinely do such a thing as part of his or her studies. It is not so hard, although twenty years ago it would have been impossible. We can add a gene or remove a gene, or we can modify a gene that way.

"In principle we can do anything. This is not to say we understand what we are doing, but the technology is very general. If you want a specific DNA molecule, you can order it on the Internet. You can type in its sequence down at the Cyber-Yak Cafe, and a company will make it for you the next day.

"Cloning is another technology, which is different from the DNA work I have been describing, but it is appropriate to consider it here. Cloning also involves the ability to go from a single cell isolated in a dish back to a complete animal. In cloning, we take a cell, remove its nucleus with the DNA, and put the nucleus into an egg cell whose own nucleus has been removed. When you put the nucleus from a mouse into a mouse egg, you can stimulate it to divide and produce a whole mouse. Dolly the sheep was the first mammal to be cloned in this way. Ten years ago this could be done for a frog, but it was not possible for a mammal.

"Aside from making an animal that way, you can also let those cells continue to divide in a dish. The hope is to treat those cells in a way that will cure a disease. For example, if I had diabetes and could not regulate my blood sugar, this would be due to the lack of certain cells in my pancreas. In this case I would like to grow such cells in a dish and inject them into myself.

"Many people in the West are worried about this, saying that the minute you have conditions for cell divisions that have the potential to become an animal, a soul exists, and it would therefore be unethical to use those cells. This is causing great consternation in America, because maybe people who are opposed to abortion would say that these cells in the dish are already an embryo. And technically they qualify. To me, they're not sentient, but of course I'm using the gross level of consciousness as my criterion.

Buddhism might recognize that a subtle consciousness must already be there if those cells have the potential of becoming an animal. It's a difficult issue.

"And here is another issue that may be even a little bit harder: suppose I have a human disease in some gene. I can produce a mouse that has that same human disease, which is inflicting cruelty on the mouse. However, making such models of disease is also our best way to come up with a therapy and gives us much more power to try to cure the disease. To my mind, this is an appropriate trade, but it causes ethical and philosophical consternation for some people. In general, how should one think about this?"

The biological manipulation of memory provided another example. "I have a friend at Princeton who has been studying memory, and he thinks there is a particular protein that plays an important role. He put extra copies of that gene into a mouse and then tested its memory, and found that the mouse had a much better memory for certain things. When the scientist removed that gene, he found that the mouse had a much worse memory for those things. No one has yet done this manipulation of memory in people, but these experiments raise the possibility. The same questions may also apply to any kind of perception.

"Cancer raises other issues. In a cancer, certain genes get broken. One could put an extra copy of that gene into the cells, making it harder for the cell to have both copies broken. Sounds good. Someone did this experiment: they put an extra copy of the gene in question into a mouse and they checked whether the mouse resisted getting cancer. It did, more so than a normal mouse. But there was a problem: it also aged prematurely.

"There is nothing simple in this work. It reminds us, of course, to be humble about how little we understand. We had this great prediction about protecting against cancer, and we were correct. But we forgot to predict that it would promote aging. We now actually think we know why it causes aging, and we probably should have thought about it, but we didn't.

"So what are the ethical choices we have? I was careful here to focus on a mouse, but in principle all of these things can be done in humans— though I don't know whether in Buddhism that distinction matters. But in practice, what should we do?

"We can, in some ways, select our children. Once a woman is pregnant, we can obtain cells from the fetus, test their DNA, and determine that this child will have a particular genetic disease. I have friends, one of them a scientist, who had a child with a rare genetic disease that causes death within nine months. The gene was not known, but that scientist worked very hard to find the gene so that in a future pregnancy, they would be able to know if their child would have the same disease.

"The same thing is possible for Alzheimer's disease. I can tell you if a fetus has a genetic 'spelling' that will make him or her much more likely, though not certain, to have Alzheimer's disease. What do we do about that? And of course, in much simpler terms, there are questions of selecting a boy or a girl, which I find abhorrent because of the lack of respect for life. But when I look at my friends who have the child with this disease that may cause him to die at nine months, I cannot blame them for wanting to avoid another such pregnancy. But what about someone who would make this choice for a fetus to avoid Alzheimer's disease? I don't know where to draw the line.

"We can make things a little bit easier, perhaps, by doing the analysis on cells in a dish instead of waiting until there is already a fetus. We can take eggs and sperm, unite them in a dish, and let several different eggs begin to divide. From each embryo that has begun to divide into multiple cells, you can pull off one cell, which is enough to check the DNA of that embryo. We might therefore see that, of eight embryos, four have the disease and four do not. We can choose which one to put into the womb."

This was a very daring experiment to propose to his audience, and Eric looked around to gauge the reaction. He saw only very attentive faces.

"What is the status of the other ones that we did not put back? What is our motivation? How do we think about that choice? Again the question arises, when we have just eight cells in a dish: Are these already souls? Is this cell already a person? If it is, how do we think about not putting back the other seven? If it's not a person, then maybe it doesn't matter. These are choices we will have to make.

"Then there are the questions around cloning. We can clone a mouse, a sheep, a goat. No one has done it with people. Personally, I hope no one ever does. I suspect it is possible, but I don't see what would be gained by doing it. This is a topic of great debate in the world today and my question is, what do we lose by doing that? Some people say that we would lose nothing; there is no harm. Others would say that we change our whole notion of human beings and our notion of children if we select their genetic make-up. And this becomes a subtle question: How will we be different as people if we start to consider our children to be products of manufacture rather than products of nature?

"This has led some people to ask again about cloning cells—not growing the whole human being, but just cells in a dish to replace, for example, pancreas cells that are not functioning properly. In the United States right now, we have a great problem with this question. Most of my scientific colleagues would say it is a very easy distinction: cloning a person is problematic, but there is no problem with growing the cells. But that is not our president's position, or that of many other well-meaning people in the United States who would say, 'No, the minute this has the potential to become a human being, it has a soul and to manipulate it is killing.' Of course, by not doing this work, we may be letting people die who would benefit from these cells.

"The last and hardest issue is the question of improving, changing, or manipulating the genetic text. We have two types of genetic texts that we can possibly change. One is the genetic text in our own body. There are people, for example, who cannot make the proteins necessary to clot their blood. In principle, one could introduce the gene to correct that into a small number of cells into their bones. That genetic change would not be passed on to the children. We call that genetic therapy of the body, or somatic genetic therapy.

"But other genetic therapies, if they affect the sperm, the eggs, or the embryo, will be passed on to every generation. Like the mouse with a gene that improved its memory, we might be able to do the same thing for a person. But it might turn out like the mouse that aged prematurely. Right now I feel that such manipulation is not safe, because we are not able to predict the outcome fully. But at some point we will know technically how to do it, and be able to evaluate the consequences. And then we must ask: Is this is wise idea? Should we? Shouldn't we?

"I don't know, but we are going to have to make a choice about that. The minute we make the choice to actually produce a child that has such a change, then our children are in some ways a product of manufacturing. And that changes our whole way of thinking. It means that some families may be able to afford 'better' children. I don't like that word, but I use it only in the sense that they may be able to afford a child who has some genes that prevent cancer. And others will not be able to afford that. But of course, even as technically proficient as we are, we still make mistakes. What do we do to take back a mistake? I don't know.

"We have many questions to talk about, but to my mind the most important lesson that I take away from genetics is a lesson of understanding and compassion." Eric closed his presentation by showing a poster from a museum in France saying Tous les mêmes, tous différents, which he translated into English as "'All the same, all different' or 'All related, all different'—we are all the same, different by only one letter in a thousand. And yet that one letter in a thousand can make a difference. This is our struggle right now, to understand how we are all the same and how we are all different. How do we understand ourselves? Do we use this knowledge to separate people or bring us together?"

Reflections

The Human Genome Project is certainly one of the main achievements of life science today. I remember the fierce controversy that accompanied the start of the project, with high waves of emotion from both scientists and the lay public. There are several aspects of the project that lend themselves to debate. The very idea that the intimate secrecy of our own human code would be revealed and made public is disturbing, and the intensity of the debate has been augmented by short-sighted statements claiming that deciphering the human code would be a panacea for many diseases afflicting humanity.

But the project can be also seen as a reductionist approach to life, reducing the whole genome to an array of single components. The Harvard geneticist R. C. Lewontin takes a general stand against genetic determinism,[2] namely the claim that all human existence is controlled by our DNA. In this light, he questions the assumptions that have been made about the Human Genome Project, and he writes, for example:

> Even if I knew the complete gene specification of every gene in an organism, I could not predict what the organism would be. Of course, the difference between lions and lambs is almost entirely a consequence of the difference in genes between them. But variations among individuals within species are a unique consequence of both genes and development environment in a constant interaction. Moreover, curiously enough, even if I knew the genes of a developing organism and the complete sequence of its environments, I could not specify the organism. . . . It is not that the whole is more than the sum of the parts. It is that the properties of the parts cannot be understood except in their context in the whole. . . . The theory that searches for human nature in the products of genes . . . misses the whole point.

Eric Lander in his presentation gave a good picture of what the Human Genome Project can do and cannot do, and he avoided any close relationship with genetic determinism. But even without genetic determinism, ethical problems are bound to arise in this field. Alison Abbott describes an interesting ethical problem related to the use of the human genome for the treatment of diseases.[3] It is known that many people are treated with drugs that, for them, have no effect. We believe this is largely due to genetic variations that affect the metabolism of the drugs, determining whether or not a drug will work or what its minimum dose should be for a certain individual. Now, "regulatory agencies such as the US Food and Drug Administration (FDA) are starting to consider whether or not some drugs should be labeled as being suitable only for individuals with a defined genetic profile." This may lead us to a world in which each individual is mapped for his or her own genes. Will we all in the near future carry our genetic profile in our pockets, along with other pieces of ID? This is just one case in which the new genetic achievements are bound to raise complex ethical questions.

It is interesting to observe that this great scientific achievement has produced a big shadow, or a series of shadows. Where do we go from here? The only way perhaps is to avoid radical, fundamentalist positions against or in favor of the Human Genome Project and analogous biotechnology projects like the use of stem cells or cloning. We should instead assume a behavior that is contextual, with each issue evaluated and discussed by the community in its particular context of scientific and ethical conditions. This will not be easy, but it may avoid bigger problems.

An Interview with His Holiness the Karmapa

It happened at the end of one of the morning sessions. Matthieu Ricard had told me to stand up immediately and follow the Karmapa right away. In a few seconds, however, he was mysteriously out of sight, and I was confronted with empty corridors and closed doors. I had lost track of Matthieu too, and of my son Peter, who was supposed to record the interview with one of those miniature electronic gadgets that only young people can master. I wandered at length from one silent hallway to the other, already certain to have missed my long-awaited appointment. Then all of a sudden Matthieu appeared in a doorway and made a sign, Peter came running from the other side, and we were all gathered in the Karmapa's beautiful private room. He was sitting patiently in a big chair.

Matthieu gave me a sign to start. The question that I had formulated was something of this sort: "Your Holiness, you have been following with great attention the words of the scientists. However, your background in physics, molecular biology, and genetics is, as you told us, almost nonexistent, and you are young. What is it that you are following so intensely? What are you learning, and what does your mind do in these circum-

The interview with His Holiness the Karmapa, with Matthieu Ricard as translator.

stances?" While I was talking, I had the impression that he was already lost in his own thoughts, and indeed, his answer at first was indirect.

"If you ask me generally why I find a pleasure in science, it is because of its affinity with Buddhism. I give you two reasons: one is that both science and Buddhism are based on a relation of cause and effect. Whatever happens is based on a cause that in turn gives an effect. In this sense I am at ease with the scientific debate.

"The second reason is that both science and Buddhism are based on a strong appreciation of validation. There is no blind faith. It is true that occasionally in Dharma there is a lot of superstition, but in general both science and Dharma rely on solid proofs. And this corresponds, actually, to my natural inclination. . . ."

The Karmapa looked at me all of a sudden, and there was a profound sense of earnestness in his eyes. At this point I tried to go back to my original question: "But how do you deal with the mass of scientific information that is reaching you?"

"I do not always appreciate the actual information," he said. "But there are two important aspects that I appreciate and on which I rely: one is the description of the external world; the other is the reality of mind. When I want to know more about a certain subject, I know that I can pursue a deeper knowledge of science and be very enriched by that."

"In your relation to science, do you plan to follow in the footsteps of His Holiness the Dalai Lama?"

There was a gentle smile of humility in his answer. "In terms of aspiration, yes. But I do not feel comfortable yet, particularly with the English terminology. I have to study English first, and then learn about the basic things of science without the barrier of language."

"Do the discoveries of modern science occasionally cause you fear?"

Again he gave me an answer full of common sense and sincerity: "On the one hand I appreciate the power of new discoveries, and I value the progress of science, particularly when it brings the well-being of people. On the other hand, yes, it is clear that these discoveries also have a negative potential."

I tried to end on a personal note. "Are you homesick for Tibet?"

His face became very serious. "I was in a tight situation there, but homeland is always a good place. More specifically, my family is very much beloved to me. This is natural," he concluded with a determined look.

6 / From Consciousness to Ethics

Dear reader . . . recall the bright, joyful eyes with which your child beams upon you when you bring him a new toy, and then let the physicist tell you that in reality nothing emerges from these eyes; in reality their only objectively detectable function is continually to be hit by and to receive light quanta. In reality, a strange reality! Something seems to be missing in it.

—ERWIN SCHRÖDINGER

THROUGHOUT THE CONFERENCE, the themes of consciousness and ethics surfaced often in the informal discussions that followed each morning's scientific presentation. We have seen these themes foreshadowed at various points, such as Michel Bitbol's contribution relating emergence and consciousness or the understanding of cellular awareness at the foundations of life that is part of Francisco Varela's legacy. Eric Lander brought to our attention many of the ethical dilemmas that have surfaced recently as a consequence of progress in the study of genetics and the human genome. Ursula Goodenough also talked about the possible biological basis for compassion and empathy, as well as links between genetics and human culture.

But the study of consciousness in Western science is still in its infancy when compared to the long tradition of contemplative exploration in Buddhism. Likewise, in the West it often seems that ethics is applied to science as an afterthought. Research speeds ahead, and only when some innovation of science is visible on the horizon, or already within reach, do we pause to consider its ethical implications. In contrast, the understanding of ethics is deeply embedded at Buddhism's core and integrated with its view of consciousness.

We will focus now more closely on the Buddhist contribution to the dialogue, beginning with an overview from Alan Wallace, which offered a rich and lively account of Buddhism as a science of the mind, its origins, and its ethical implications.

A view of the whole conference hall, waiting for His Holiness to arrive.

Buddhism as a Science of the Mind

"Over the last few days we've heard wonderful, synthetic, and systematic presentations from one field after another within science," Alan began. "In the dialogues we've had little bolts of light coming in from Buddhism, but very sporadically and in a very unorganized way. And so, very briefly, I'd like to give the big picture from the Buddhist view, something systematic and comprehensive that might facilitate further dialogue, though with nothing of the detail of the scientific presentations.

"Steven Chu mentioned that the history of empirical science begins at the time of Galileo, which is 400 years ago. But it's actually more than 2,400 years since Democritus said that atoms moving in space are all that really exists. So these ideas have a long history, although empirical science and its rapid accumulation of knowledge is 400 years old. It's only in the last 120 years that the West has had an empirical science of the mind. In other words, we ignored the mind for close to 300 years before it entered the field of science.

"But even then, researchers rapidly shifted to focusing on behavior, which brought them back into familiar territory. They finally came around to attending to the mind in cognitive psychology, but it is always somebody else's mind. And by the time we get to neuroscience, it's somebody else's brain. So the mind has been reintroduced, but usually on familiar territory, either studying behavior or studying the brain.

"Introspection, or first-person observation of our own mental processes, which after all is the only immediate access we have to any mental process, remains at the level of folk psychology in the West right now. There is no penetrating, rigorous, reliable, cumulative science of first-person observation, experimentation, and refinement of the mind. William James, the American psychologist who founded the first experimental laboratory of psychology in the United States, suggested three strategies for understanding the mind: exploring the mind through behavior, through brain science, and through first-person introspective investigation. The first two are familiar. We are getting better and better at behavioral studies, and we have tremendous sophistication in neuroscience. Introspection seems not to have moved."

After this rousing introduction, Alan let his hands rest a few seconds, which meant that he also stopped speaking and probably thinking. First-person science is a subject he feels passionately about, and a theme of his book *The Taboo of Subjectivity*. He continued: "Questions such as what happens to the mind at death, when exactly mind emerges in the formation of the embryo, the nature of mind and of consciousness—these remain, for scientists, metaphysical questions. And so it's worth pausing to think whether the mind and consciousness are intrinsically metaphysical in nature. Or is it simply that we have not found the right tools for developing a systematic, rigorous science of mind and consciousness in terms of the phenomena themselves?

"Empirical science leaped forward when there was sufficient technology for people like Galileo to make very close observations of phenomena and see patterns; as better technology developed, there was better penetration of the phenomena. Science didn't start out with just a bunch of assumptions, but with careful firsthand observation and experimentation on the phenomena. That has not occurred with the mind. So I raise the question once again: Are mind, consciousness, and questions like what happens to the mind at death simply unscientific, or is it a matter of cultural orientation? Did we, with our Greco-Roman, Judeo-Christian background, simply focus on pragmatic, empirical questions that look outward to the physical? Might it be that another civilization, every bit as intelligent and sophisticated as ours, chose to ask other questions?

Another scene during the conference.

"Democritus articulated his outward-looking view, that finally every-thing consists of atoms, around 2,400 years ago, about a century after the Buddha left his home in search of truth."

The Buddha's Story

Here Alan gave us an account, with his own modern inflection, of the story of the Buddha's search for enlightenment. "When the Buddha set out from home, he joined a very loose-knit movement of individual wandering ascetics or seekers who were rebelling against, or at least deeply skeptical of, the prevailing Vedic religion. The Vedic religion was very dogmatic and heavily ritualistic, asserting that salvation was to be found only by per-forming a prescribed set of rituals very accurately. People like Gautama, who would become the Buddha, and many of his peers, were revolutionar-ies, independently seeking.

"The first thing he did was to seek out two of the greatest adepts, ac-complished masters of this counterculture, and he trained with them in some of the best contemplative technology that had been devised. At that

The reclining Buddha.

time, they had already achieved extremely subtle, focused states of concentration, known as *samadhi*. Gautama was clearly a prodigy with a genius for this training, because in a very short time he achieved the same degree of samadhi as his teachers.

"But he was not satisfied, because no profound, irreversible transformation had taken place. When you come out of samadhi, you're essentially back in the same state you were before, still vulnerable to suffering. In other words, it's like a contemplative vacation: a very nice holiday, but then it's back to the treadmill. The Buddha was really looking for a way to sever the very roots of dissatisfaction. He acknowledged that he could find nobody who knew how to do that, so for six years he did his own experiments. He looked at different types of experimental strategies: of diet, behavior, exercise, types of mental processing. He was a wild experimenter and went through a lot of suffering. He earned the great respect of his peers because he was so utterly dedicated.

"Finally, after going from extreme to extreme, he sat down and saw that his body was weakened from extreme asceticism. Then someone came along and asked if would like a nice, simple meal of rice and yogurt. *Maybe a good idea*, he thought. And so he had a good meal, and then sat down with great determination, saying, 'Now my physical health is restored,'— so in a way he was acknowledging the importance of the body's relation to the mind—'and now I will seek.'

"He now used the method of samadhi that he had trained in earlier, but instead of taking that state as an end in itself, with its attendant bliss or great peace, he used samadhi as a tool for active investigation. His primary area of interest was not, as in the case of Galileo, whether a ball rolling down a ramp will accelerate or move at a constant velocity, or whether the Earth goes around the Sun or vice versa. His primary area of interest was just the opposite of the West. His questions centered around the nature of the mind.

"Using the laser-sharp focus of samadhi, he penetrated through what we call the ordinary mind—the mind that is busy remembering, imagining, and desiring, *I want this, I don't want that.* He calmed that level of mind so that it became completely transparent, and he was able to tap into a deeper stratum of consciousness. Being very curious, he directed that very pure, focused attention back through his own life, and then back to previous lives, which manifested to him with complete clarity. Tapping into deep memory, he was able to recall an endless number of preceding lives and the circumstances of each life. As far back as he looked, he could never find a beginning. So he stopped looking, effectively saying, 'I have enough data.'

"He then expanded his vision, attending to other people's mind streams, and he saw how their mind streams, just like his own, seemed to recede endlessly. Then he did what we might call a meta-analysis, examining whether, in this stream of lifetimes, the behavior of one lifetime is related coherently and causally to the events of a later lifetime. He could see patterns: if this happens, then that happens; where this does not happen, that does not happen. He didn't speak of an underlying mechanism but rather of a phenomenological causality. From these observations, he came to certain principles or laws concerning *samsara,* which is what we call the cycling through one process of becoming after another, over and over again.

"All this took place in one night. It was a very good night. Finally he raised the question: What can bring samsara to an end? How can we free ourselves from being thrown compulsively and involuntarily from lifetime to lifetime? By morning, he saw how to accomplish the cessation of this cycle. He abandoned the very causes that perpetuate that cycle, and his mind was completely freed of all obscurations.

"From a Western perspective we would ask: What god gave him that knowledge? Was he a prophet, or was he simply a great philosopher? From the Buddhist perspective, he was not a prophet. He was very clear that this knowledge was not given to him by somebody else, nor did he just think it up, nor did he get it from the surrounding society. He said it came from an empirical, rational, pragmatic, and skeptical mode of inquiry. It was empirical because he was observing, looking directly. It was rational because he made sense of it and organized it into a theory that is internally consistent and consistent with available knowledge. It was pragmatic because his fundamental motivation was to identify the roots of suffering and eradicate those roots for himself as well as for others—so it was also compassionate. And he held skepticism about the prevailing ideas, rituals, creeds, and beliefs of his era.

"So that makes him sound like a remarkable, one-of-a-kind individual. And at first he wondered if anybody would understand what he had to offer. He paused and considered, noting that a few people would have 'a little dust on their eyes.' A few people would get it. So he went out and taught, and basically, all of his teachings for the next forty-five years were structured on these four basic issues.

"Suffering is a reality. Rather than avoiding or denying it, we should recognize it and explore it in its entirety. Does it have causes? What are those causes? We should check empirically and find out what is crucial. There are many contributing factors for suffering, but which of those factors that are within our control—within the mind—can be eliminated? Is it possible to irreversibly remove those tendencies of the mind so that suffering never comes back, even though outside circumstances are terrible? And if so, how? Finally, there is a strategy to accomplish this.

"During the 45 years that the Buddha taught, he encouraged his followers to put these teachings to the test, not to accept them out of reverence for him. You must check them, he said, just as if you were buying gold: you would check it by burning, rubbing, cutting, and only when you were completely satisfied would you then purchase it. He encouraged skepticism. For the last 2,500 years, or 100 generations, an unbroken lineage of Buddhist contemplatives has used a set of experiments and methods for making careful observations, to try to rediscover these truths for themselves."

Ethics as the Foundation

"The basic line of practice is actually very simple. The foundation of all Buddhist practice, and without which there is no Buddhist practice, is ethics. Ethics, in Buddhism, begins with the understanding that we are not just observers in reality. Whether we like it or not, we participate, with our bodies, our speech, and our minds. Certain behaviors of body, speech, and mind are injurious, whether in solitude or with others. They lead to our own unhappiness, to conflict, and to the suffering of others. Much of physical suffering comes from nature and is beyond our control. But the suffering that human beings inflict on others is within our control. What mental processes give rise to human-made conflict, misery, and suffering? What behaviors of body, speech, and mind bring benefit: greater peace, happiness, and relaxation, and also greater harmony with others? This empirical, rational, pragmatic approach is the foundation of Buddhist ethics.

"For contemplatives who follow the Buddhist path to become buddhas themselves, rather than merely studying Buddhism, ethics takes on the further meaning of refining the mind. It becomes part of their contemplative technology. In the West, one problem with introspection, which has had a very short lifespan within the scientific context, is that the tool used was very poorly trained. The ordinary mind that oscillates between excited agitation and dull laxity is not a good instrument for making observations. This problem was recognized by the Buddha, who adopted existing techniques, brought them into Buddhism, and applied them in uniquely Buddhist ways. Without ethics—if you're behaving out of anger, craving, or delusion, and thereby reinforcing these mental afflictions—meditation can make no progress. A very fine sense of ethics is necessary for developing this contemplative technology of a mind that can enable you to make careful, precise, and deep investigations.

"The final phase of Buddhist practice is the cultivation of insight, where you explore the world as it exists, not independently of experience and consciousness, but rather the world of experience called in German *Lebenswelt*, or in Sanskrit *loka*. The strategy for this begins with trying to discover the nature of the objects of consciousness, the objects of mind. If I gave you a tool and said, 'Look through this tool and you'll find many interesting things,' as a good scientist, you would ask, 'What is the tool?' You would want to take it apart to understand how good a tool it is.

"Before investigating the rest of reality, you have to examine the tool that is the mind. Scientists waited three hundred years before examining this tool, and there's still no science of consciousness. How can you have a science of consciousness when you have no scientific means of exploring the nature of consciousness?

"The scientific community is like a living organism. Science in 2002 is not the same as the science of 1802 or 1602. Like a marvelous living organism, science has been drawing in nutrients, excreting waste products, and mutating in various unexpected ways through history, economics, war, and a myriad other influences. It evolves but maintains its integrity. It has never stopped being science. It's growing now and it seems quite healthy.

But as Steven Chu pointed out, the power and knowledge of science is so great that if it's used with hatred or malice, it could destroy humanity. Science urgently needs to absorb the nutrients of greater wisdom and compassion for its own self-preservation, if for no other reason.

"Buddhism is also like a living organism that has evolved. We do not practice exactly the same Buddhism now that the Buddha revealed 2,500 years ago, but there is a coherent identity. You can recognize that this really is Buddhism. Traditional Asian forms of Buddhism have now entered the very different environment of the modern world. How do we preserve its integrity? It has to adapt, but if it adapts too much, the organism dies and is replaced by little fragments. It could become little more than a therapy, helpful but very far removed from what the Buddha had in mind. On the other hand, if we try to preserve too much, it will also die. If it is seen as irrelevant, no one will feed it.

"Both of these living organisms have to adapt. If science doesn't adapt, it may destroy all of us. If Buddhism doesn't adapt, it may just die. But the two may come together in a form of symbiotic mutation. They already share four basic elements of empiricism, rationalism, pragmatism, and skepticism. Maybe emergent properties will arise, new ways of inquiry that preserve both scientific and Buddhist values."

A Dialogue on Consciousness

Alan Wallace's overview presented a clear picture of Buddhism as a science of mind, of the importance of the notion of consciousness in Buddhism, and the neglect of that idea by Western science. From the opening day of the conference, consciousness became an important theme in the afternoon discussions that followed each morning's scientific presentation. These discussions were free-flowing and engaged the entire group. Questions might be sparked by the earlier presentations, but new avenues of inquiry often opened spontaneously, as we will see.

Arthur Zajonc opened the discussion that day by posing a question to the Dalai Lama. Reminding us of Matthieu's earlier presentation of the Buddhist concept of a beginningless universe, he extended this notion: "How does Buddhism account for the arising of life and sentient beings? Is this also beginningless? Luigi described the very gradual development of life through increasingly complex organization and the emergence of higher properties. Is there an analogous description in Buddhism concerning the nature and emergence of life?"

His Holiness responded by first qualifying the concept of beginningless-ness. "If one focuses on a specific locale and a specific time frame, then it is possible, of course, to speak of a relative beginning, such as the begin-ning of this planet. So we have no problem with the scientific description there. Likewise, if we look at the origins of life in terms of sentient beings who have corporal bodies made out of matter, we accept the scientific ac-count on the whole and we learn a lot from that. But if the question con-cerns the origins of an individual sentient being, as opposed to the body that the sentient being possesses, then the issue becomes much more com-plicated. The origins of consciousness are also much more complex."

I sensed that we were approaching a critical divide between science and Buddhism. I thought we might as well face it directly, so I asked, "Does Your Holiness accept the view we have heard, which is also Francisco Va-rela's view, that consciousness arises naturally as an emergent property at a certain level of brain and neuronal complexity?"

"It's very clear," the Dalai Lama answered, "that specific modes of em-bodied consciousness, such as the human psyche or human visual percep-tion, do not arise in the absence of the brain or the appropriate faculty. The brain definitely contributes to the emergence of visual perception and vari-ous aspects of the psyche. But if we examine the clear, luminous, and cog-nizant aspect of these mental processes—in other words, consciousness itself—then the Buddhist perspective is that the event of consciousness does not emerge from the brain or from matter."

"This is an important difference," I noted. "Many scientists accept the idea that all properties of mankind come from within, even consciousness and the idea of God, as self-generated values. This is not so for Buddhism."

"That's correct," the Dalai Lama affirmed.

"There is an issue of terminology here," Alan Wallace said, clarifying a point that had been missed in translation. "His Holiness spoke of gross and subtle levels of consciousness. Gross consciousness is comprised of those processes of the human psyche, like visual perception, that are con-tingent upon the body, the brain, and the nervous system. Subtle con-sciousness, which carries on from lifetime to lifetime, has no beginning, and is not dependent on the body."

The Dalai Lama continued with his explanation: "If we compare a plant, for example, with the human body, there is a great deal of common ground in terms of how the cells are organized, as you well know. But does a plant have any recognition or experience of good and bad, of pleasure and pain? Does a plant have conscious experience? You won't find an explanation for that simply by understanding the processes within the cells. Consciousness doesn't emerge from the cells. Consciousness only arises from consciousness. It does not arise from matter.

"A gross level of consciousness, such as visual perception, is contingent upon the body and depends on the mechanisms of the retina and the visual cortex, as well as on the objective referent, whether that is photons or the object that you are seeing. But in the midst of the visual perception there is the very element of awareness, which is called *rigpa*. This is the luminous experiential component, the cognizant component of visual perception, or of knowing. We're speaking now of a subtler level, which does not arise from the visual faculty or from the objective referent. It arises only from a previous continuum of consciousness."

At this point, His Holiness threw us a challenge that seemed to come straight from the debating ground of a Tibetan monastery. It was rigorously logical, but still foreign to us. Summarizing very briefly the scientific view of the origin of matter and of evolution, he homed in on two points of divergence in this history: "Some configurations of matter in this process of cosmic evolution provided a basis for the emergence of life and some did not. What was it about the configurations of matter that enabled them to become a basis for life, and enabled that to become a basis for consciousness? What is it that distinguishes organic from inorganic matter? Within the organic track, what differentiates the organization of the cells that provide the basis for consciousness from those that do not, such as plant life?"

I was nonplussed by the question, coming as it did on the heels of my presentation on self-organization, complexity, and emergence. Had I not been understood? "This is straight Darwinian evolution," I said. "Once you have life, it evolves in very different directions depending upon environment and accidents of contingency."

"But what is unique to matter in your earliest animal that is conscious?" the Dalai Lama persisted. "What special property of its matter enables it to be conscious, whereas its predecessors, and other tracks of evolution such as plants, are not?"

"Through contingency and evolution," I answered, "the structure of the brain evolved in such a way that permitted the arising of the capability for self-reflective awareness. Plants evolved in such a way that this capability never happened to arise. The straight answer would be the neuronal complexity of the brain in one case, and not the other."

It was Alan Wallace who was translating for His Holiness at this point, and he seemed to be enjoying the confrontational flavor of the debate.

Thupten Jinpa stepping in with a smile.

"What aspect of the complexity? What type of complexity? Complexity doesn't explain anything!"

Here Thupten Jinpa stepped in, as if to calm the waters. "The explanation is really based on the degree of complexity."

"Yes," I said gratefully, "and in order to have consciousness, according to the traditional science view, you need billions of neurons, which you do not find in a fly or in a plant." I was frustrated by the way we seemed to be talking at cross-purposes, and happy when Eric Lander signaled that he would like to contribute, as he is well-versed in evolutionary science and had proven himself a good communicator. But the bluntness of his statement surprised me.

"In fact, we don't know," Eric admitted. "We have no idea. Consciousness is not a property of matter, because the matter in my body will change over the course of a year and will come from the water and elsewhere. As scientists we think it is a property somehow of the organization, but we have no idea. What's interesting to me is that Buddhists, as Matthieu explained, seem very disturbed by the idea of a first cause for the universe. I share that disturbance, which is not to say I am any happier with the idea of beginninglessness. That also disturbs me. But you go from the idea that there is no first cause for the whole universe to the idea that there can be no first cause for consciousness. It seems to me that in Buddhism you can't imagine consciousness arising from nothing. And scientists, perhaps because of our worldview, cannot imagine a different explanation. I don't know that either of us has a logical reason to say that it must have persisted forever or that it must have arisen from complexity. In science we have so little to say about it because so few experiments try to probe consciousness. Mostly we avoid the question."

Steven Chu stepped in here, first affirming what Eric had said—"The simple answer is, we don't know"—but then offering an explanation for the Western bias favoring the notion of life as an emergent property. Western science, he said, has had much success with the notion of emergent properties: from basic chemistry to recent work in superconductivity, superfluidity, and lasers, we find examples of surprising collective phenomena. "Every physicist would agree," Steven said, "that there's a much bigger jump from atoms, molecules, cells, and neurons to consciousness. We don't pretend consciousness is comparable, but perhaps small successes at a much simpler level have made us overconfident."

Michel Bitbol had a very different view of the matter. "I would like to correct the impression that there is a wide gap between the Western and Buddhist views of consciousness," he said. "Nowadays Western philosophers have good reasons to criticize very strongly the idea that consciousness is an emergent property of a complex chunk of matter. One is that science, by its very method, is unable to grasp what consciousness is. In order to make good predictions, science has to exclude everything that is not common to everybody. For instance, when science speaks about heat, it excludes the felt quality of heat, and only retains what can be measured with a thermometer and shared on a piece of paper. Since it excludes felt qualities from the outset, present methods allow no possibility of explaining them. A second reason is that consciousness cannot be called a property, nor even a phenomenon. A property is something objective, something detached from us, which is attributed to an object, whereas consciousness obviously cannot be detached from itself. Likewise, consciousness is not something that can be considered a phenomenon. Rather, it's phenomenality by itself. For these two reasons at least, and there are many others, many philosophers nowadays think that consciousness must be considered primary, and not derived from anything else."

Arthur Zajonc turned the question back to the Dalai Lama: "When we think of consciousness arising from matter, we do so because it is hard for us to imagine it any other way. What is the alternative? Could we imagine a universe where consciousness somehow exists while matter is still in a very rudimentary state? You seem to imply that it's possible to have sentience or consciousness of some kind without the bodily support of a complex physical organization. Do you have any empirical evidence for that?"

His Holiness answered, "Even in Buddhism there is an implicit recognition of the difficulty of identifying what consciousness is. Although we are aware that consciousness exists, when we try to define it, it becomes very nebulous and difficult to pinpoint. But in principle, Buddhism maintains that it is possible to recognize experientially what consciousness is and identify it. There is an understanding, for example, that a highly advanced practitioner at the point of death can identify something called the 'clear light of death,' which is regarded as the subtlest experience of consciousness.

"The issue of consciousness is indeed difficult to explain. I encourage those of you scientists who are studying what we call grosser levels of consciousness—forms of emotion, the neural correlates, and so forth—to keep up the good work. And on the Buddhist side, meditators have to continue working very hard to achieve higher levels of consciousness. In ten or twenty years, with more meditation, a more convincing, truthful discussion can take place."

Ursula Goodenough raised a question that would lead to a helpful insight: "Does Buddhism recognize a relationship between the human experience of consciousness and the mental experiences of a chimpanzee, or an organism with whom we clearly share a common ancestor?"

I was reminded how often in past conferences the Dalai Lama himself had posed questions of comparison between humans and other animals to clarify a concept. Here he answered by mentioning first that Tibetan folk mythology traces the origins of human beings to monkeys, although, he said, classical Buddhist texts suggest that human beings devolved from more subtle forms of body, composed of light, with characteristics very different from normal human beings. "But according to Buddhism, many of the basic emotions, such as altruism, compassion, and greed, are felt by animals just as in human beings. There is only a difference of degree of complexity."

Having found some common ground, Ursula pressed further. "Is there a sense that these emotions or states are transfigured in humans by our subtle consciousness? Do we experience them differently from animals because we have this other property of consciousness? One view in science is that humans access the primate mind, but transfigure it so that it becomes enriched or more abstract, able to be organized in new ways, because we have this other property of consciousness."

"There is a difference between animals and humans in how these emotions are experienced, in terms of complexity and probably the degree of self-consciousness," His Holiness answered. "But in the Buddhist view, the difference between the animal realm and the human lies more in the level of intelligence. As for subtle consciousness, there is no difference between animals and human beings. Any sentient being that has the capacity to experience pain and pleasure is thought to possess this subtle consciousness."

So the subtle consciousness is not uniquely human, we learned, and if we had been tempted to equate it with Western notions of the soul, here was a deep divergence. In all this I had not yet heard a clear refutation of the notion of consciousness as an emergent property, so I now restated my original question to His Holiness: "How would you refute the idea that consciousness is a self-generated value that arises as an emergent property from the self-organization of the structure? Could you accept this as a possibility?"

The Dalai Lama answered, "If you are asking whether it is possible for matter—the brain, our cells, and so forth—to act as cooperative conditions that mold or influence the gross manifestations of consciousness, such as visual perception, the answer is yes. But then we ask what consciousness actually arises from. What is it that turns into consciousness? According to

P. L. Luisi asking whether consciousness isn't an emergent property of the brain.

Buddhist principles, consciousness can arise only from a continuum of phenomena similar to itself, in the same way that formations of mass-energy give rise to formations of mass-energy. It is a similar continuum. Subtle consciousness is a radically different type of phenomenon; therefore it can arise only from phenomena similar to itself. Matter, configurations of mass-energy, is radically dissimilar to consciousness. Therefore only a stream of consciousness can give rise to a later stream of consciousness. Matter cannot transform into or become consciousness.

"I fear that science sets up a false dichotomy when you have separate streams: inorganic matter giving rise to life forms, finally giving rise to consciousness. And then, quite correctly, you question the Buddhist view of consciousness giving rise to life. But it's not true that Buddhism views consciousness as the substantial cause out of which matter emerges. If A is the substantial cause of B, then A actually transforms into B, and in so doing it loses its earlier identity.

"If Buddhism adopts the notion of the big bang as the beginning of this universe, then the origin of matter in this universe is not a preceding continuum of consciousness, or divine consciousness. Nothing like that. The substantial cause of the first matter in this universe was preceding matter. Only mass-energy gives rise to mass-energy, and consciousness always gives rise only to consciousness."

"Isn't that a very dualistic view?" I asked.

"Yes, it's a form of dualism," His Holiness responded. "The moment there is more than one thing out there, dualism arises. There are all kinds of dualism. For example, one of your dualisms is inorganic and organic matter. Dualism makes sense only in relation to a very specific context.

After a brief pause, he continued. "Of course, we do make categories. For example, we distinguish between permanent phenomena and impermanent phenomena, and within the category of impermanent phenomena we make distinctions between mental phenomena, abstract composites, and physical phenomena, or matter. Not everything falls into the category of matter. Perhaps if we were fully enlightened we might see things differently."

Eric Lander asked a question that challenged the Buddhists on their own terms, wanting to know the ethical and practical implications of what might otherwise be seen as an obscure point of philosophy and cosmology. "The idea that subtle consciousness does not arise from matter and is not dependent on matter is clearly a central point to Buddhism. Your Holiness says that science cannot disprove this point, and I agree. As I listen, though, it seems to me that this idea in Buddhism is more an unproven but accepted assumption than it is a conclusion resulting from proof based on evidence or logic. Suppose that someday you concluded, which we cannot do today, that the subtle consciousness *did* arise from matter. Would it change anything? How would a Buddhist act differently?"

"First of all," His Holiness began, "it's not true that this is merely an assumption. There's an empirical basis that is repeatable. There is a systematic training that can lead to the empirical conclusion that a continuity of consciousness transcends the limitations of one body, one life. This is not something unique to Buddhism; it preceded Buddhism, and it is not embedded in one ideology or one belief system. There are different modes of meditation within Tibetan Buddhism, different avenues to that experience. In the cultivation of samadhi, for instance, you train in highly refining and developing the mind to such a degree of subtleness, clarity, and stillness that you can penetrate through the turbulence of your gross human mind, with all of its sensory perceptions and all of its cogitations. You settle your rapidly churning mind so that it becomes transparent, and then you tap into an underlying core of memory, including memories of previous lives that lend themselves to objective corroboration.

"There's another mode of meditation within Tibetan Buddhism, called Dzogchen, where you do not necessarily have to achieve a deep state of samadhi, but you let your awareness come to rest in a very subtle way, where again the turbulence of the gross mind is calmed. Occasionally during this state something arises, like a bubble coming out of water, where one remembers a previous life, or a hundred lives, just like that"—here he snapped his fingers—"even as far back as the Buddha's time, or before that. I have not had that kind of experience, but some of my friends have.

"There have also been small children who, without spiritual training, have somehow accessed very clear memories of a past life. I've actually met a four-year-old Indian girl who remembered her past life so clearly— her home, her parents, their name, their professions, the place they lived—that the parents of her previous life were fully convinced that this girl was their own child."

Here the Dalai Lama conferred briefly with the translators before offering some further thoughts that demonstrated both Buddhism's way of pursuing a logical argument rigorously to the end, however rarified its destination might seem to Western thinking, and its down-to-earth pragmatism. "In addition to this," he said, "Buddhist philosophy employs the logical reasoning that if consciousness can arise from matter, then we have to posit a beginning to consciousness and a beginning to the continuum of sentient beings. By extension of that reasoning, we would also have to accept a beginning to the whole universe, which opens up a whole can of worms. Since Buddhism rejects that and accepts the beginningless continuum of consciousness, therefore it also accepts the beginningless continuum of sentient beings. Because sentient beings have no beginning, Buddhism interprets the evolution of the physical universe as intimately interdependent with the sentient beings who inhabit and experience the external world.

"As to the question of why it matters, first, it presents a philosophical problem. If we are forced to accept a beginning to the universe, we have two options. Either something comes from nothing, or we have to posit a divine creator, a transcendent being, neither of which Buddhism finds comfortable. Second, from a soteriological point of view, a single lifetime is an extremely brief duration in which to achieve liberation and enlightenment. It's said to be possible in principle to achieve enlightenment in three years, three months, and three days, but this is much like Communist propaganda. The chances of it happening are so remote you might as well forget about it! Even in a lifetime of sixty years, the chances of achieving enlightenment for most of us are remote. So we need a bit more time."

Matthieu Ricard offered some further clarification: "I think the proper description of subtle consciousness might be pure consciousness, or the most basic quality of consciousness. Different emotions are, of course, already colored by mental activity; but what is common to all mental activity, and what allows it, is this most basic faculty that we call luminosity, which is not colored by thoughts, concepts, or reasoning and yet is different from blank storage. It is simply the most fundamental, pure quality of consciousness. And the law of cause and effect applies to that, which is why we need a continuum. Maybe it's not a question of billions of years, but of just two consecutive instants—but the cause has to be in harmony with the result. In just the same way that right now, a grain of rice will not produce wheat in one single transformation, the very preceding instant of consciousness, which relies upon that basic luminosity, has to be of the same quality. Whatever the mental coloring might be, there has to be a continuum of this primordial clarity because the very nature of consciousness is not qualified by mental efforts. And therefore there is a beginningless continuum."

Steven Chu interrupted, speaking with intense sincerity. "I'm listening to this and I understand some of the words, but I feel it's like a physicist explaining electromagnetic waves to someone who doesn't know mathematics. We try very hard and we use analogies, but in the end you need years, even decades, of study before you understand. I get glimmers of what you are saying, but I haven't had the years of training and the discipline to really fully understand." There were probably a few others in the same boat with Steven. Matthieu listened carefully, and continued.

"We've been speaking about the first-person and third-person experience in science. I think it's important to distinguish different degrees of first-person experience. Perception, for instance of colors or of heat, is a first-person experience that you cannot replace with any equation. Most of science—all the equations, all the numbers—is third-person experience. But a more fundamental first-person experience is that pure consciousness, pure luminosity, which is not even qualified by mental processes. That is the ultimate first-person experience, which, as Michel was saying, is a primary quality."

"And that takes years of study," Steven chimed in.

"And practice," Arthur added.

Alan Wallace pointed out that meditation practice resembles physics or mathematics more than many other disciplines: "You can't learn physics or mathematics just by reading a lot of books or listening to people talk. You have to do it yourself. You can read all about meditation, but if you don't do it, you won't understand it."

"Is everyone capable of doing it?" Eric Lander asked.

"Is everyone capable of having your knowledge of the genome or of mathematics?" Alan countered. "This is similar. People come in with different capacities. Some will be better at it, some will be worse at it. But everybody has the capacity to improve."

The Dalai Lama added a characteristic note of humility: "We need to recognize our current limitations in how far we can claim to be able to describe the nature of reality. We might conceivably study all three hundred volumes of the Buddhist canon and still could not claim to know every aspect of consciousness and all phenomena that are part of mental consciousness.

"Similarly, scientific knowledge is limited, and the fact that we cannot observe particular phenomena with current scientific methods cannot be taken as grounds to claim they do not exist. The limitations of scientific knowledge quite clearly leave open the possibility that there may be other types of physical phenomena that have not yet been discovered. They also don't preclude the possibility that science might one day access some nonmaterial phenomena. I'm quite sure that in time the scope of scientific research will expand, especially in terms of investigations about the brain, emotion, and intelligence. I think that is the proper way to learn more

about consciousness. More experiments are needed, including collabora-
tions with experienced Buddhist practitioners."

Consciousness and Emergence

We had returned again and again to questions about consciousness and
emergence in the discussions throughout the week. On that final day,
Arthur Zajonc was expressing all his concern: Were we approaching a
very sophisticated, subtle understanding of how higher-order properties
emerge? Or was our use of the term "emergence" problematic, perhaps
facile? Earlier, Michel Bitbol had warned of the dangers of an ontological
understanding of emergence, steering us on a middle path between the
fallacy of the emergence of intrinsic properties and the equally mislead-
ing idea of emergent properties as mere epiphenomena. Was there some
way, Arthur asked, that we could be more careful in our use of the term
"emergence"?

Alan Wallace responded with a contribution in his own voice, removing
his translator's hat for the time being. "From a Buddhist perspective, there
seems to be an explanatory gap from cellular awareness to brain-based
awareness. We have this complex configuration of neurons, synapses, and
neurotransmitters in the brain, all of which can be studied objectively. We
assume we'll understand them more and more comprehensively. At some
point, science implies, we might even discover consciousness.

"But what is consciousness? Let's take a simple example: the perception
of the color blue. I'm looking at Barry's shirt now, and I'm perceiving blue.
We understand much of how that takes place, in terms of photons reaching the retina and signals going back to the visual cortex. It's certainly possible, if it hasn't already happened, that exact one-to-one neural correlates
could be found for the perception of blue.

"So now we are presented with two kinds of phenomena: an emergent
property of that complex configuration of brain activity that does indeed
arise from more basic elements, and the actual subjective experience of seeing blue. Here's the explanatory gap. In all other cases, physical properties
emerge from physical properties that emerge from physical properties . . .
and all of those can be seen in the physical world. But in the case of perception—here, the subjective experience of the perception of blue—there's a
gap. You can see the exact neural correlates associated with that perception,
but they're not blue. Those neurons are just the same color they always were.
There are no yellow neurons or red neurons, any more than there are red

photons or yellow photons. To actually experience the perception of blue, you have to look out of your *own* eyes and see something blue.

"This appears to be a radically different type of emergence, if, in fact, this is emergence. It's profoundly different. No matter how carefully you scrutinize the brain, you never see blue, and no matter how carefully you scrutinize blue, you never see the brain. Whereas if you scrutinize any other emergent property, you can see what it emerges from. If you look at the wetness of water, you can see water. If you look closely at water, you can see molecules. If you look closely at molecules, you can see quarks. It's like having one big microscope. But if you look at the brain, there's no microscope that will show you blue, or any other object of perception."

I had to express a reservation at this point. "Alan, treating the perception of blue as a kind of emergent property is confusing and unnecessary. The machinery of perception might be considered an emergent property, but the blue is a result of this emergent property. Likewise, we consider consciousness as an emergent property, but your feelings of fear or love are not emergent properties. They are results or consequences of this particular faculty."

Alan stopped, but then decided to pursue his train of thought instead of being diverted by my objection, and continued, "Although it's a reasonable hypothesis that subjective, brain-based awareness is an emergent property of the brain, it's clearly not a scientifically established fact at this point. It's a cogent theory, but only a theory, so we can at least consider other theories. The Buddhist theory is that our subjective awareness is not brain-based awareness, but rather brain-conditioned awareness. Human consciousness is indeed an emergent property, but it emerges from a deeper level of consciousness, one that carries on from lifetime to lifetime, and is then conditioned by the brain.

"For example, fifty-some years ago at my conception, the stream of consciousness that came in was not a human consciousness. It was not Alan Wallace's consciousness. It wasn't a human soul. It was simply a stream of consciousness that had an enormous amount of experience behind it. When it came in, it set up initial conditions in my mother's womb that would determine the type of person I would become. Of course, there were many other conditions contributing as well. But this human psyche, heavily conditioned by brain, body, environment, parenting, education, and so forth, actually emerges not from the brain itself but from something similar to itself. That is, it emerges from a stream of consciousness, just as configurations of mass and energy emerge from earlier configurations of mass and energy, despite how radically different they may appear." It was now time for Alan to make another jump.

Testing Reincarnation

"This is a theory, just as the scientific view is a theory. One theory suggests that at death, the brain decomposes and no longer functions as a brain, so the emergent properties of brain-based consciousness just vanish. The Buddhist theory is that, since Alan Wallace's consciousness did not emerge from the brain, that consciousness will dissolve back into the source when the brain ceases, and the continuum of mental consciousness will carry on. Both theories are compatible with neuroscience. Neither one contradicts what we actually know of the brain and the mind.

"So then we can ask how this could be tested scientifically. The theory from science, that death leads to absolute cessation, is very difficult to validate or repudiate. If we can't even imagine how to disprove it, I would question whether it is truly a scientific theory. (A third theory available in the West is the Christian theory that at death the human soul goes to heaven or hell or limbo or purgatory, but it doesn't come back. We have no way to test this scientifically either.) The Buddhist theory says there is continuity: the person will come back again, and in some cases—especially those with very high training—that person will come back with clear, accurate memories of the preceding life. Of the three theories, that's the only one we can test."

Eric Lander inserted a qualification here. "Just because you can't find a way to test a prediction doesn't mean it's not a scientific theory."

Alan responded by offering an alternative way to validate the scientific, materialist theory: "If all of the necessary and sufficient causes for consciousness could be determined, just as we have defined the minimally necessary conditions for life; if we could demonstrate that consciousness doesn't arise in the absence of those causes; and moreover, if we could show that all those necessary and sufficient causes are physical, this would be very compelling proof of the materialist theory. But for that, you would have to have an objective means of detecting the presence or absence of consciousness in anything. There is no such instrument. There may be something like that in the future, but at present there is no instrument that can test whether a rock, a butterfly, an amoeba, or a human being is actually conscious or not. All we have to go by is behavior, which can be simulated by unconscious robots and computers.

"To test the Buddhist theory scientifically, you would need a systematic series of studies, not simply the testimony of a four-year-old girl here and a Tibetan *tulku* there. You would need a contemplative laboratory, where people trained rigorously for maybe ten or twenty years, and then you could test their memories very rigorously. That would open up some very interesting collaborative research between science and Buddhism."

"I want to make a small but important point," Steven Chu interjected. "In science there is no absolute proof. We can never prove an idea or theory, even with hundreds or thousands of experiments. If you meet a person who claims to remember his last life, you cannot say, 'Ah, then it must be so.' It means that the person feels they remember their life, and this is only supporting evidence, and each instance is further supporting evidence, never a proof. Similarly, each experiment on atoms is only supporting evidence, never a proof. If my students say, 'This proves it,' I say, 'Wash your mouth out.'"

Michel Bitbol answered Steven, "What Alan tried to explain is that the only possible confirming evidence about the content of consciousness is first-person, or subjective evidence. No other type of evidence can be obtained. The only real difference between the two sides, I think, is the type of evidence that is relevant for an analysis of consciousness."

"No," Eric Lander objected, "he's actually suggesting the use of second-person evidence. He's meeting science on its own grounds. If one were to demonstrate many examples of accurate memory of past lives, he would argue that this falsifies the theory that consciousness does not continue. We could disagree about whether such evidence is adequate—we could have many, many problems with this evidence, and we do. But I respect the fact that he's actually laid out a test." Steven agreed that this kind of experiment is in principle worth doing, and many voices chimed in, all of them positive and every scientist stressing the importance of rigorous standards.

Arthur Zajonc pointed out that some detective work on verifying past-life memories had already been done by Western psychologists, notably Ian Stevenson. This prompted some reflections from Matthieu Ricard on the cultural bias that has prejudiced science against investigations of the thorny question of reincarnation. "Ian Stevenson studied six hundred cases over a period of thirty years," Matthieu pointed out, "and he is very well accredited. He discovered that most cases did not prove anything—they might have been fake or inconsistent—but there were twenty cases for which he could find no reasonable explanation other than memory from past lives. So there were facts and large scientific publications, but there is definitely a cultural resistance. Those books are hardly examined critically.

"This is a very important question," Matthieu continued. "Imagine what kind of change it would make in our perception of life if we truly had a clear indication of our death. I remember participating in a TV program in France about the SETI Project, the search for extraterrestrial life. In France, this subject is taboo, and the whole program was closed down. A major French scientist who was instrumental in closing it said in front of five million people, 'If there were a green man landing in my garden, I would just close my windows and continue my work.' That was an extreme position, but there is a similar resistance against taking reincarnation seriously."

Ethics and the Analysis of Reality

It was during these richly textured afternoon discussions that the Buddhist side of our dialogue was heard most clearly. Having begun this chapter with Alan Wallace's account of the origins of Buddhist ethics in the Buddha's own life story, we will bring it full circle with another remarkable presentation on the philosophical grounding of Buddhist ethics. It was the final session of our meeting and the mood was relaxed, a mix of satisfied exhaustion, camaraderie at the completion of the long week's journey of the mind, and a keen awareness of this very special time and place to which we would soon be saying good-bye. Matthieu Ricard spoke with a gentle fire of conviction.

"Earlier I mentioned the relation between a correct perception of reality and ethical issues—how we can implement proper understanding in order to lead our lives and transform ourselves. We might ask whether it is because we are Buddhist practitioners that we want to relate those two, or how the relation between our understanding of the perception of reality flows naturally into personal transformation, because it may seem artificial to link those two things together. So I thought I should say a few words about the intimate relation between those two. This actually bridges the scientific approach to describing phenomena with the need, whether Buddhist or human, to transform ourselves into better human beings and actualize our potential.

"When we analyze reality, we have a tendency to solidify or reify phenomena. Although we know that things are impermanent, that they are always flowing, that nothing remains identical to itself even for two consecutive moments, yet we have this tendency to perceive that yesterday's table is today's table, that the person we meet today is more or less the same person as yesterday. Even more deeply, although we know that we change from youth to old age, we think there is something constant that is 'me.' Instead of seeing the fluidity of phenomena, whether external or within our minds, we grasp them as being solid. What was fluid water now becomes ice.

"What are the consequences of that? Instead of perceiving the intimate interdependence of constantly changing phenomena and understanding that nothing can happen except through relationship, we instinctively try to ascribe intrinsic properties to things. The first great divide, of course, is between self and others. What is in truth completely interconnected becomes two worlds: me and all the rest of the world. Then from 'me' comes 'mine': 'This is mine; that is not mine'—my relatives, my belongings. Then we start to ascribe properties to things and people. We say, 'This is beautiful.' Somehow we cannot help but feel that this beauty, or that pleasant aspect, intrinsically belongs to this person or that object. Little by little, we solidify, crystallize, and divide everything as being pleasant, unpleasant, beautiful, ugly, delicious, disgusting, mine, others'. We have passions, impulses attracting us to what is pleasant to me, what I like, what I want to attract, to get, to keep, to increase. And we have a tendency to repulse whatever causes fear or disgust or animosity, because we ascribe those feelings as intrinsic properties. This is a friend, this is an enemy; and both become solid identifications.

"What comes next is a very big development. Our mind is invaded by a chain of thoughts that arises from those feelings of attraction and repulsion. Attraction becomes a strong craving, an obsession, a desire that completely invades our minds. We feel pride and superiority when our self relates to others we consider inferior. We feel jealousy when something we consider ours is taken away. We feel animosity, wanting to destroy or harm what seems threatening to us or goes against our desires. We lack discernment because we are blinded by this host of emotions and toxic mental events. We can't discern with correct judgment what needs to be accomplished and what needs to be avoided in order to fulfill our most intimate wish, the longing for happiness, and the longing that other beings have for happiness.

"All these different metatoxins become the way our mind functions, to different degrees. Sometimes the mind is stronger than other times, but in the end all this leads to a deep feeling of frustration and suffering because the world will never match our desires. So there is a very close relation between our first misapprehension of the nature of phenomena—finding solid, intrinsic properties in an increasingly fragmented vision of the world—and suffering.

"If you turn this around, you perceive interdependence. Instead of building reality out of separate, permanent, intrinsic qualities, there is a whole dynamic flow of relationship, constantly in transformation. It crystallizes in different ways under different conditions, according to your perception and so many other factors that you cannot isolate individual causes as pleasant or unpleasant. This vast net of interconnection is described in the *sutra* as a necklace of pearls on the palace of Indra, each pearl reflecting the whole palace. If you perceive things like that, then the whole process of solidification will not happen. You'll naturally have the understanding that things are impermanent and changing, that the enemy of yesterday can be the best friend of tomorrow, that what seems beautiful to someone seems ugly to someone else, that you yourself are changing from one minute to the next."

Realizing Interdependence

"You will also see that there's no such thing as the constant entity you perceive within yourself, the 'me,' if you really analyze it. In this stream of constantly changing consciousness, there is no permanent boat that is the self. In realizing this, instead of losing what you might feel is the most precious thing in yourself, the 'me,' you are not losing anything. You are just unmasking an impostor. It is not the most precious thing in your being, but rather, it is what ties you to suffering.

"Realizing the interdependence and the dynamic flow gives you freedom. You're no longer driven by the mechanism of taking and rejecting in the same way. There's no longer a reason for strong animosity to arise from identifying someone as truly, intrinsically an enemy. Ice and water are of the same basic nature. But the ice that comes from solidifying phenomena can cut. You can break your bones on it. If it melts, through inner freedom from the solidification of concepts, it's just fluid. It no longer threatens to harm your happiness. Losing the self does not mean becoming nothing; you simply untie the knots of solidified phenomena and gain inner freedom.

"This also has a very intimate link with compassion. Compassion without interdependence means nothing. Our true nature is love. Imagine yourself suspended in space. If there were no interdependence with other beings, then there's no notion of love or compassion. Understanding interdependence makes you understand that your happiness comes through others' happiness. There is no way you can build your happiness at the

A break for tea and further interesting discussion.

cost of others' suffering. You might gain a temporary satisfaction at having defeated your enemy, but that will never be a lasting happiness.

"For these reasons, a correct understanding of reality—the absence of any intrinsic nature of phenomena, and their interdependence—is said to be the ultimate view of the Buddhist teachings, referred to as wisdom. And this is intimately linked with compassion, love, and altruism, which are the expression of this understanding and the quintessence of Buddhist ethics or behavior. Wisdom and compassion are like a bird's two wings. A bird cannot fly with one wing . . ." (The Dalai Lama giggled to himself, then shared his joke: "The bird would fly in circles!") "And you cannot ask which wing should start first," Matthieu continued. "Both work together. You cannot start to fly with the right wing only and get to the left one later. You won't fly very far that way. We have to keep wisdom and compassion in union all the time, from beginning to end, uniting understanding with ethical thoughts, words, and actions."

At Matthieu's conclusion, there was a spontaneous outburst of applause throughout the room, and the Dalai Lama expressed the general sentiment in a single word. "Excellent!" he said.

Reflections

There was a profound silence after these words. And then, in a few seconds, it was as if a couple of things rapidly crystallized out of this silence.

One was the statement by His Holiness that consciousness is a very complex business, difficult to grasp and define. It sounded as if the Dalai Lama himself could not offer a clear, final word on the issue. He may not have intended it, but to me his statement was at once sobering and frustrating.

Another element that now appeared crystal clear and was likewise not entirely happy was the apparent impossibility of reconciling the scientific and Buddhist views on the nature of the subtle consciousness. On the one hand, there was a strong statement from the Dalai Lama, as well as Matthieu Ricard and Alan Wallace, that subtle consciousness has no material basis. Plain and simple: it does not arise from the material world. The scientists had extreme difficulty with this, insisting, although in different words, on questioning why it should not be an emergent property of a sophisticated form of mental organization. We scientists were all uneasy with the Buddhist view, which was tantamount to introducing a transcendent element. Science, in its most traditional definition, deals with the phenomenology of nature, trying to explain it in terms of the laws of nature—the laws of physics, chemistry, and biology—and even though the limits of science may stretch because these laws are continually being modified and expanded, there is no room for transcendental elements. Is there a way to reconcile these two positions? In the silence that fell over the room, they appeared to me as two separate, crystalline truths.

As a partial help, the words of His Holiness came to my mind, when he said (in the preface to Daniel Goleman's *Destructive Emotions*) that we should discriminate between what science finds to be nonexistent and what science does not find. What science finds nonexistent, he said, we should all accept as nonexistent. But when science does not find something, it is a different situation, and consciousness is one such case.

It was interesting to me, given the scientific opposition to transcendence, that it seemed we all knew and accepted this argument, as if to say, "Well, okay, we will keep the door open a crack. . . ."

7 / Last Words

AS THE FINAL AFTERNOON of our week together drew to a close, Arthur Zajonc asked each of the participants to speak for just a minute or two about the most significant impressions that we would carry home from our dialogue. What had we learned, what had we gained by being here?

Eric Lander spoke first. "I knew nothing about Buddhism when I came here, and I've learned a tremendous amount about the basis in ethics, the belief in the importance of relieving pain and suffering, and the impor-

The group of Western participants at the Mind and Life conference.

tance of logic and reason, evidence and debate. In many ways this is similar to what motivates scientists. It's different in some ways, such as the sort of evidence we accept, but I admire the similarities, as well as the importance of challenging authority in both traditions. In many ways it is different from many Western religious traditions.

"Another thing that struck me is the importance of motivation. That is not something that usually figures in Western debate. In fact, we have a saying that implies motivation is *not* important: 'The road to hell is paved with good intentions.' I was also struck by the importance of middle paths in Buddhism, the importance of compassion, and Your Holiness's understanding of choices that others would make, even if they were not the choices that you yourself would make.

"I know that Your Holiness and the monks knew more about genetics at the beginning of this week than I knew about Buddhism, but I hope you still learned something about genetics. I hope that perhaps some of the younger monks, or the older monks, will write about the ethical questions in genetics. Right now in the West, people are debating these ethical questions, and I hope that Your Holiness and the monks here don't feel that you have to wait until you know all of the scientific details. I would be very interested in your perspectives, particularly on the issue of motivation and how it enters into the writings, the struggle, and the debate. Right now motivation is not a large part of that debate. I know that what is going on now is far too extreme, with too much rejection and too little of the compassion that you hold to be so important. Several of the monks have said, 'Well, I have so much to learn.' You also have so much to teach. Thank you."

It was my turn next. "I've been at all the previous Mind and Life meetings in Dharamsala, and one thing that struck me particularly is the much deeper dialogue between science and Buddhism in this meeting. In previous meetings, Buddhism and science were winking at each other, friendly, occasionally touching, but they were still respectfully distant. I felt that in this meeting there was a real engagement, not just a flirtation anymore.

"Also, I ask myself whether this congruency brought me closer to Buddhism. Perhaps yes, but as Matthieu said, reincarnation—and the accompanying notion of nonmaterial-based subtle consciousness—is a dividing line and it still remains a dividing line for me. Put lover of Buddhism on one side and Buddhist on the other, depending on whether or not they have this faith. I know this is not what you think, but I have to say what I think.

"The congruency between Buddhism and science also gave some support to one of my major lines of reasoning and study—Francisco Varela's congruency between the organic structure and cognition, and the fact that you cannot have consciousness without the human being, and vice versa—the notion of embodied consciousness.

"The final point, again personal, was the feeling of being one of the more privileged people in the world, privileged to be in this place of beauty among people who are so advanced in knowledge. I often feel a sad note together with this sense of beauty and privilege, a questioning: 'Do you deserve it? Or are you playing a trick on your karma?' I don't know whether I am playing tricks on karma, but I am very happy to have met many new good friends in this meeting. I will enjoy these new friends for the rest of my life, and I am very thankful for that."

Steven Chu then offered humbly, "I have to interpret things I have felt in this simple mind in simple ways. It's like quantum mechanics, where you can only keep track of eight atoms or else it's hopeless. I have to say first that earlier we had a very beautiful discussion with many of the monks on the reality of a tree. During this discussion I was wondering why they are so interested in the observation that the reality of the tree is not really a reality, that it's part of our perception and everything else. As the week went on, I began to realize that there was a reason for this, a purpose for this worldview. I realized it's not just simply a matter of being connected with the rest of the world. You have to be so connected that you can say an obvious thing like, 'If you're good to people, they'll be good to you; if you're mean to people, they'll be mean to you. And so it's better to be good.' But it goes much deeper than that. As Matthieu said, you shouldn't even separate good people and bad, but rather just take this as a whole. You shouldn't make any distinction like that, because in the end it's destructive on an emotional and personal level. Originally, these ideas were way above my pay scale, but now I appreciate the deeper meaning of how we get to this happiness that we all want. This, I believe, has deep truth. There may be no evidence, but I believe it.

"When I gave my presentation, I talked about how physics tells us that there are certain properties at the most fundamental level—measurements of the spin, the position, and the momentum of particles—which are deeply connected with the rest of the world. But there are other properties like charge where we don't see that connection, and I was amused that all the properties don't necessarily have to be connected. If only some of the properties are indivisibly connected, this is enough.

"There are limits to our knowledge. We may never understand things that are too small or too big, and we may never understand things that are too complex, because they are connected. Physics has success where things are weakly connected. When things are strongly connected—and life is

The monks at the Mind and Life conference.

strongly connected—then it becomes very difficult. But I really mean it when I say you don't have to be fully connected to be connected."

Michel Bitbol then took his turn. "Let me tell you how I perceive this meeting now that it's over. This conference is like no other I have attended until now. There was no question of using ideas to enhance the egos of the speakers, or to further their careers. Instead, I found that among us the ideas, instead of being shot at one another, had a highly significant cooperative value. Even if we didn't renounce our ideas, we are now changed in a subtle way. Something truly collective has emerged out of this. It is due, of course, to Your Holiness's extreme attention, respect, and openness, which catalyzed everything. In this way, to use Matthieu's metaphor, our ideas have melted. They are really helping one another by their patient and deeply sympathetic confrontation. Out of the cacophony has emerged a choir, though with persistently distinct voices, a choir conducted by an exceptional tuner of minds."

Matthieu Ricard was the next to speak. "At this completion I should probably apologize for having brought up the subject of reincarnation. It is an aggressive concept to Western mentality. In fact, my deep conviction is, as His Holiness said, that it doesn't matter if the universe has a beginning or not. We have this consciousness anyway; we experience it. The main point is how to contribute to alleviating suffering and bringing more well-being for ourselves and for others. That's the very reason you came out of your laboratories and I came out of my hermitage. So let's fortify that fruit. This is my hope, prayer, and wish."

It was Ursula Goodenough's turn. "Being last, [I find that] everyone else has already said what I was going to say. Before I came here, Buddhism was an abstract idea for me, something that some people were interested in and some weren't. The most important result is that Buddhism is now real to me. It's now you people. It's now a practice, a way of being. It's the students down in the yard debating. It's the people in the temple and the wonderful chanting. It's Alan and Jinpa, and the Karmapa and Matthieu with their presence. And I will never be the same. I thank you so much."

Arthur Zajonc added a few final words. "We set out to address the questions What is matter? and What is life?. The exchange has been far richer than I've experienced before in such dialogues, from both the side of Western science and the side of Buddhist introspective inquiry. For that I rejoice. I'm very pleased at the content, the quality, and the seriousness of our exchange. But beyond the specifics of the content was the character and style of that exchange. There's an extraordinary energy and a great joy, a humor and engagement that, to me, epitomized what I'd always hoped being an academic would be like. To come here and do that together has been a really great privilege.

His Holiness the Karmapa and his satisfied look.

"In some ways, the most important theme of the meeting began with Your Holiness's opening remarks about the basis and purpose of this meeting: connecting to each other not only through our knowledge but also through our values as human beings. This theme has come back again and again throughout our meeting like a leitmotif, and it has two sides. On the one side it came up as a topic for discussion, but from the other side, I felt it as a reality. From both Buddhism and Western science, there are people present here with good hearts. We bring that to our conversation and in doing so, we not only answer questions about matter and life, we discover each other. I am also very, very grateful for that."

And finally, our host the Dalai Lama brought the circle to a close: "First of all I would like to express my deep appreciation to all of you, and particularly to the scientists who have taken the trouble to carve this time out of their very busy professional lives to make this opportunity for us to meet here, especially those who come from the States. To come from the United States to Dharamsala is not an easy task.

"Our discussion has been very informal. Each of you expressed what you feel, and exchanged views and feelings as if among old friends. I think that's important. After all, we are all visiting together on this small planet. I think differences are secondary, whether based on nationality or race or ideology or system. The most important thing is that we are human beings.

"I want a happy, meaningful life, and everybody has a similar desire. I think it's unrealistic to expect something from outside to make our lives happier or more meaningful, so whether one is happy or has a meaningful life must depend on oneself. The environment is very important, but ultimately the key factor is oneself. The atmosphere we show to each other as human beings is essential for a happy, successful, and meaningful life. As a Buddhist, I always pray that some space remains for me to contribute to happiness for the infinite number of other sentient beings. That's my own share, and I consider this meeting part of that. So I think it is important that more progress has been made in our meeting, and this progress should and will continue. We need to make effort on both sides, Buddhists as well as our friends here.

"As individuals, each of us will eventually be gone. This applies to me too, of course. I am sixty-seven years old, and my individual life is limited. We have already elected the next Tibetan leader, which means I am now preparing for another life. If there's no next life, that's okay, it's all much easier. If there is a next life, I have to prepare for it.

"Generation after generation, each must develop, and then must go. Therefore we need more effort now to carry on this work, and I hope that eventually more will develop from this. Until now, it has mainly been Westerners and Tibetans working together, but eventually people from other nations, like China, should join us. I really wish for us to become more and more compassionate, and I hope we will also be joined by people from other Buddhist traditions in China, Sri Lanka, and Thailand.

"Once more, I really appreciate your contributions. As a Tibetan, in our tradition, once we develop a friendship, that friendship will remain until our death. So I will remember all of you. Thank you."

As tired as we were, most of us would have loved to continue talking and listening for hours more. We were all aware that the material presented that afternoon could have filled another whole week, and perhaps would keep us working for an entire life. As we filed out of the room, the sun was still high and bright, and we all felt keenly the privilege of being where we were.

About the Mind and Life Institute

R. Adam Engle and Zara Houshmand

THE MIND AND LIFE dialogues between His Holiness the Dalai Lama and Western scientists were initiated in 1984 by R. Adam Engle, a North American businessman, and Francisco J. Varela, a Chilean-born neuroscientist then working in Paris. Both were Buddhist practitioners and aware of the Dalai Lama's desire to deepen his understanding of Western science, in which he held a long-standing and keen interest, and to share Eastern contemplative knowledge with Western scientists. Engle and Varela each independently conceived of a series of cross-cultural meetings where the Dalai Lama and scientists from the West would engage in extended discussions over a period of days. Joan Halifax, then director of the Ojai Foundation, urged Engle and Varela to combined their efforts. Michael Sautman and the Dalai Lama's youngest brother, Tendzin Choegyal Ngari Rinpoche, also played instrumental roles in the early efforts to organize the dialogues.

The Dalai Lama welcomed the initial proposal with an enthusiasm that he has sustained, along with a significant investment of his time, over two decades. After long planning, the first meeting was held in October 1987 in Dharamsala. It was decided to focus on the scientific disciplines dealing with mind and life as the most fruitful interface between science and the Buddhist tradition, and this also provided a name for the Mind and Life Institute when it was formally established in 1990. Engle and Varela collaborated closely to develop a useful structure for the meetings. Engle took on the job of general coordinator, primarily responsible for fund raising and organizational aspects, and has served as chairman of the institute since its inception. Varela, as scientific coordinator, was primarily responsible for the scientific content of the early meetings, inviting scientists, and preparing a book about each meeting. The scientific coordinators for later meetings were chosen for their expertise in the varying topics, but Varela remained a guiding force in the Mind and Life Institute until his death in 2001.

Cross-cultural dialogue is notoriously difficult to engineer, but the Mind and Life conferences are unique in the care that has been devoted to building bridges that can mutually enrich modern science and the Buddhist tradition. In an effort to avoid the pitfalls of earlier encounters, Varela urged the adoption of several operating principles that have worked extremely well for the Mind and Life dialogues. Perhaps the most important is that scientists are chosen not solely for their reputations and competence in their domain but also for their open-mindedness. To ensure that the meetings will be fully participatory, they are structured with presentations by Western scientists in the morning sessions, designed to brief the Dalai Lama on fundamental scientific knowledge from a broad, mainstream, nonpartisan point of view. The afternoon sessions are devoted solely to discussion, which naturally flows from the morning presentation. During the discussions, participants can present personal preferences and judgments, if they differ from the generally accepted viewpoints.

The issue of Tibetan-English language translation in a scientific meeting posed a significant challenge, as it was impossible to find a single individual fluent in both languages and familiar with the terminology of both science and Buddhist philosophy. This challenge was overcome by choosing two highly skilled interpreters who work side by side during the meetings to provide instantaneous clarification of terms and are able to identify cultural stumbling blocks and add insight from their own substantial knowledge of the two domains. Dr. Thupten Jinpa, a Tibetan scholar who is the Dalai Lama's primary translator, holding a Geshe degree from Ganden Shartse monastery and a Ph.D. in philosophy from Cambridge University, has participated in all of the Mind and Life dialogues. Dr. Alan Wallace, a former monk in the Tibetan tradition with a degree in physics from Amherst and a Ph.D. in religious studies at Stanford University, has interpreted for most of the dialogues. Dr. Jose Cabezon and Dr. John Dunne have translated when Dr. Wallace was not available, and the Venerable Matthieu Ricard has also contributed his skills as an interpreter at many meetings.

A final principle that has supported the success of the Mind and Life dialogues has been that the meetings were entirely private in their original format. The absence of press and the very limited number of invited observers has offered a protected environment to conduct this exploration. Recently, in response to the Dalai Lama's request to share the dialogue with a wider public, several meetings have been open to press and to scientific audiences as well as to Tibetan scholars from the monastic community. The Mind and Life Institute makes archival recordings of the meetings and prepares the proceedings of each for publication as a book, with DVD recordings also available for some of the more recent meetings.

* * *

In 2000, at Mind and Life VIII on Destructive Emotions, the Dalai Lama observed that it appeared to him that Buddhism and other contemplative traditions had developed practices for training the mind that cultivated a number of positive mental qualities that reduce suffering and promote health and well-being. He asked the scientists assembled in the room to please test these contemplative practices in their laboratories to determine if they are, in fact, helpful, beneficial, and useful. Thus, scientific research was added to the mission of the institute.

The Mind and Life Institute has become a world leader in pioneering this cross-cultural scientific dialogue and rigorous collaborative research between scientists and contemplatives on the effects of meditation on basic brain mechanisms, attention, emotional balance, kindness, and the prevention and treatment of disease. MLI has developed a comprehensive strategy to catalyze the creation of new fields of scientific research and academic endeavor that will support and carry out this long-term research and development agenda. We describe these new fields as:

contemplative neuroscience—the study of the effects of contemplative practices and purposeful mental training on the human brain and behavior;

contemplative clinical science—the study of the effects of contemplative practices and purposeful mental training on the prevention and treatment of disease and on the processes that underlie health and illness;

contemplative studies—the study and integration of the scientific research of contemplative practices on human brain and behavior into the existing academic fields of religious studies, theology, philosophy, and humanities.

The components of the comprehensive strategy for these new fields are:

Mind and Life Dialogues with the Dalai Lama, which occur annually in several formats and are targeted toward examining new areas of research to be explored;

Mind and Life publications, which report the content of the Mind and Life Dialogues to the general public;

The Mind and Life Summer Research Institute (MLSRI), a week-long residential "science retreat" for scientific researchers, contemplative scholars/practitioners, and philosophers to review the latest findings in the field, plan new research directions, and determine how best to advance the new areas of academic endeavor necessary to sustain the collaborative research effort among contemplatives and scientists; and the development of mental training tools based on this scientific understanding;

Mind and Life/Francisco J. Varela Research Awards, to provide seed research grants to graduate students and postdocs to examine the hypotheses developed at the MLSRI;

Mind and Life Senior Investigator Awards, for scientific research into the mechanisms and modalities of how contemplative practices affect brain and behavior;

Mind and Life Contemplative Wisdom Fellowships, for research into how the scientific data emerging from the study of contemplative practices influence the study, understanding, and teaching of religious studies, theology, philosophy, and humanities;

Mind and Life Education Research Network (MLERN), to determine how best to cultivate the mental and emotional qualities of clarity and attention, calmness and emotional balance, compassion and kindness, and gratitude and happiness among children, earlier in life and especially as part of their formal education;

Mind and Life Neuroplasticity of Self-Identification Research Project, to examine the neural systems that underlie and are responsible for how we identify with ourselves.

A Twenty-One-Year History of Accomplishment

Mind and Life Dialogues

The titles of these dialogues between the Dalai Lama and leading scientists show the range of topics that the Mind and Life Institute has explored. For more details, please go to www.mindandlife.org.

2008: Investigating the Mind-Body Connection, hosted by the Mayo Clinic

2007: Mindfulness, Compassion, and the Treatment of Depression; cosponsored by Emory University

2007: The Universe in a Single Atom

2005: Investigating the Mind: The Science and Clinical Applications of Meditation, cosponsored by Johns Hopkins Medical University and Georgetown Medical Center

2004: Neuroplasticity: The Neuronal Substrates of Learning and Transformation

2003: Investigating the Mind: Exchanges Between Buddhism and the Biobehavioral Sciences on How the Mind Works, cosponsored by Massachusetts Institute of Technology

2002: The Nature of Matter; The Nature of Life

2001: Transformations of Mind, Brain, and Emotion, at the University of Wisconsin

2000: Destructive Emotions

1998: Epistemological Questions in Quantum Physics and Eastern Contemplative Sciences, at Innsbruck University

1997: The New Physics and Cosmology

1995: Altruism, Ethics, and Compassion

1992: Sleeping, Dreaming, and Dying

1990: Emotions and Health

1989: Dialogues Between Buddhism and the Neurosciences

1987: Dialogues Between Buddhism and the Cognitive Sciences

Mind and Life Books and DVD Sets

The following books and DVD sets describe discussions between the Dalai Lama and Western scientists. Books in print can be obtained from major booksellers; DVDs are available directly from the Mind and Life Institute. For more information, please go to www.mindandlife.org.

The Science of a Compassionate Life, DVD from the Dalai Lama's Denver Public Talk in 2006

The Science and Clinical Applications of Meditation, DVD from Mind and Life XIII in 2005

Highlights from Investigating the Mind 2005: The Science and Clinical Applications of Meditation, DVD

Train Your Mind; Change Your Brain, from Mind and Life XII in 2004 (Random House, 2007)

Investigating the Mind, DVD from Mind and Life XI in 2003

The Dalai Lama at MIT, from Mind and Life XI in 2003 (Harvard University Press, 2006)

Mind and Life: Discussions with the Dalai Lama on the Nature of Reality, from Mind and Life X in 2002 (Columbia University Press, 2009)

Destructive Emotions: A Scientific Dialogue with the Dalai Lama, from Mind and Life VIII in 2002 (Bantam, 2003)

The New Physics and Cosmology: Dialogues with the Dalai Lama, from Mind and Life VI in 1997 (Oxford University Press, 2004)

Visions of Compassion: Western Scientists and Tibetan Buddhists, from Mind and Life V in 1995 (Oxford University Press, 2002)

Sleeping, Dreaming, and Dying: An Exploration of Consciousness with the Dalai Lama, from Mind and Life IV in 1992 (Wisdom, 1997)

Healing Emotions: Conversations with the Dalai Lama on Mindfulness, Emotions, and Health, from Mind and Life III in 1990 (Shambhala, 1997)

Consciousness at the Crossroads: Conversations with the Dalai Lama on Brain Science and Buddhism, from Mind and Life II in 1989 (Snow Lion, 1999)

Gentle Bridges: Conversations with the Dalai Lama on the Sciences of Mind, from Mind and Life I in 1987 (Shambhala, 1992)

Mind and Life Research Initiatives

Mind and Life Summer Research Institute—A week-long residential science retreat for 185 scientists, clinicians, contemplative scholars/practitioners, and philosophers from around the world, working together to develop new fields of science and studies that examine the effects of contemplative practice and mental training on the brain, behavior, philosophy, religious studies, and the humanities. This annual program was begun in June 2004.

Mind and Life Francisco J. Varela Research Awards—Provides seed research grants to graduate students and postdocs to investigate hypotheses developed at the ML Summer Research Institute. Ten to fifteen Varela Awards have been given yearly since 2004.

Mind and Life Education Research Network—Explores how to bring the benefits of mental training in attention, emotion, awareness, and kindness to children.

Neuroplasticity of Self-Identification Research Studies—Explores the neural systems that are responsible for the processing of self-identification, and the neuroplasticity of those systems.

Notes

1. How Real Are the Elementary Particles?

1. The conference actually started with the presentation by Luisi on the origin of life, and Steven Chu's presentation on physics came later in the week. For this book, it was decided to rearrange the material, beginning with elementary particles and following the increase in complexity to matter, the origin of life, genes, the human genome, consciousness, and ethics.
2. Although the bus full of passengers stopped in traffic is a delightful metaphor for the frozen atom, its simplification leans on classical physics and skirts the issues of quantum mechanics, where the particle "passengers" lack individual identities and well-defined trajectories.
3. Citation from *The Internet Encyclopedia of Philosophy* (http://www.utm.edu/researchiep/1/lucretiu.htm).
4. E. Schrödinger, "L'image actuelle de la matière," in *Gesammelte abhandlungen* (Vienna: Friedrich Viewweg & Sohn, 1984), 4:507.
5. Random House, 2005.
6. Alfonso Verdu, *Early Buddhist Philosophy* (Delhi: Motilal Banarsidass, 1985).
7. Hirakawa Akira, *A History of Indian Buddhism* (Delhi: Motilal Barnarsidass, 1990).
8. Reported in Charles Seife, "Physics Goes Where Einstein Sneered to Tread," *Science* 299 (Jan. 2003): 185.
9. B. Alan Wallace, *Choosing Reality* (Ithaca, NY: Snow Lion, 1996).
10. See Christian Lindner, *Nagarjuniana* (Delhi: Motilal Banarsidass, 1990).
11. See Immanuel Kant, *Methaphysical Foundations of Natural Sciences* (New York: Bobbs Merrill, 1970), preface.

2. The Emergence of Complexity

1. See, for example, Chris G. Langton, "Computation at the Edge of Chaos: Phase Transitions and Emergent Computation," *Physica* D42 (1990): 12–37; Pier Bak, Chris Tang, and

Kaspar Wisenfeld, "Self-Organized Criticality," *Physical Review* A38 (1988): 364–74; Robert C. Hilborn, *Chaos and Nonlinear Dynamics* (Oxford: Oxford University Press, 1994); Gregoire Nicolis and Ilya Prigogine, *Self-Organization in Nonequilibrium Systems* (New York: Wiley, 1977); Steven H. Strogatz, *Nonlinear Dynamics and Chaos, with Applications* (New York: Perseus, 1994); Edwin D. de Jong and Bart G. de Boer, "Dynamical Systems, Individual-Based Modeling, and Self-Organization," in *Artificial Intelligence* section in *Encyclopedia of Life Support Systems* (EOLSS), ed. J. N. Kok, developed under the auspices of UNESCO (Oxford: Eolss Publishers, 2004).

2. Ilya Prigogine and Robert Lefever, "Symmetry Breaking Instability in Dissipative Systems," *Journal of Chemical Physics* 48 (1968): 1695–70.

3. For a review of the Belousouv–Zabotinski reaction, see Richard J. Field, "A Reaction Periodic in Space and Time," *Journal of Chemistry Education* 49 (1972): 308–11.

4. See for example Theodoros Christidis, "Probabilistic Causality and Irreversibility: Heraclitus and Prigogine," in *Between Chance and Choice,* ed. Harald Atmanspacher and Robert Bishop (London: Ingram, 2002); and Steven H. Strogatz, *Nonlinear Dynamics and Chaos* (New York: Perseus, 1994).

5. For an extended literature on emergentism, see for example Pier Luigi Luisi, "Emergence in Chemistry: Chemistry as the Embodiment of Emergence," *Foundations of Chemistry* 4 (2002): 183–200, or Luisi, *The Emergence of Life: From Chemical Origins to Synthetic Biology* (Cambridge: Cambridge University Press, 2006).

6. Bernhard Poerksen, *The Certainty of Uncertainty* (London: Ingram, 2004); for more in-depth discussion about these concepts within cognitive science, see for example the books by Antonio Damasio, *The Feeling of What Happens* (New York: Harcourt, 1999); John LeDoux, *Synaptic Self: How Our Brain Becomes Who We Are* (London: Viking, 2001); and Francisco Varela, *Ethical Know-how: Action, Wisdom, and Cognition* (Stanford: Stanford University Press, 1999).

7. Francisco Varela, *El fenomeno de la vida* (Santiago, Chile: Dolmen Ensayo, 2000).

3. Toward the Complexity of Life

1. Christian de Duve, *Life Evolving: Molecules, Mind and Meaning* (Oxford: Oxford University Press, 2002).

2. Stephen J. Gould, *Wonderful Life: The Burgess Shale and the Nature of History* (London: Penguin, 1991).

3. Jacques Monod, *Chance and Necessity* (New York: Knopf, 1971).

4. Francisco Varela and Humberto Maturana, *The Tree of Knowledge* (Boston: Shambhala, 1998). For a review of autopoiesis see also P. L. Luisi, "Autopoiesis: A Review and Reappraisal," *Naturwissenschaften* 90 (2003): 49–59.

5. Francisco Varela, *Principles of Biological Autonomy* (North Holland: Elsevier, 1979).

6. Francisco J. Varela, Humberto R. Maturana, and Ricardo B. Uribe, *Biosystem* 5 (1974): 187–96.

7. Lynn Margulis and Doron Sagan, *What Is Life?* (London: Weidenfeld and Nicholson, 1995).

8. Niklas Luhmann, *Soziale Systeme. Grundriß einer allgemeinen Theorie* (Berlin: Suhrkamp, 1984).

9. Pier Luigi Luisi, *The Emergence of Life: From Chemical Origins to Synthetic Biology* (Cambridge: Cambridge University Press, 2006).

10. Humberto Maturana and Francisco Varela, *Autopoiesis and Cognition: The Realization of the Living* (Amsterdam: Reidel, 1980).

11. Francesco Varela, *El Fenomeno Vida* (Santiago, Chile: Dolmen Ensayo, 2000).

12. Humberto Maturana and Francisco Varela, *The Tree of Knowledge* (Boston: Shambala, 1998).

13. Varela Francisco, Evan Thompson, and Eleanore Rosch, *The Embodied Mind* (Cambridge: MIT Press, 1991).

14. Matthieu Ricard, Olivier Föllmi, and Danielle Föllmi, *Buddhist Himalayas* (New York: Abrams, 2002).

15. Pier Luigi Luisi, Francesca Ferri, and Pasquale Stano, "Approaches to Semisynthetic Minimal Cells: A Review," *Naturwissenschaften* 93 (2006): 1–13.

16. Ludwig Wittgenstein, *On Certainty* (London: Basil Blackwell, 1974), 33e.

17. Jurgen Schröder, "Emergence: Nondeducibility or Downward Causation?" *Phil. Q.* 48 (1998) 434–52

18. Evan Thompson and Francisco J. Varela, "Radical Embodiment," *Trends in Cognitive Sciences* 5, no. 10 (2001): 418–25.

19. John D. Barrow, *Science and Ultimate Reality: Quantum Theory, Cosmology, and Complexity* (Cambridge: Cambridge University Press, 2004); also see *The Book of Nothing: Vacuums, Voids, and the Latest Ideas about the Origins of the Universe* (New York: Vintage, 2002).

20. Leonard Süsskind, *The Cosmic Landscape: String Theory and the Illusion of Intelligent Design* (New York: Little, Brown, 2003).

5. The Magic of the Human Genome and Its Ethical Problems

1. Carina Dennis, "The Rough Guide to the Genome," *Nature* 425 (Oct. 2003): 758–59.

2. Richard C. Lewontin, *The Doctrine of DNA: Biology as Ideology* (New York: Penguin Science, 1993).

3. Alison Abbott, "With Your Genes? Take One of These, Three Times a Day," *Nature* 425 (Oct. 2003): 760–62.

Index